Reduced DC-link Capacitance AC Motor Drives

AF729257

Gaolin Wang · Nannan Zhao ·
Guoqiang Zhang · Dianguo Xu

Reduced DC-link Capacitance AC Motor Drives

Springer

Gaolin Wang
Harbin Institute of Technology
Harbin, Heilongjiang, China

Guoqiang Zhang
Harbin Institute of Technology
Harbin, Heilongjiang, China

Nannan Zhao
Harbin Institute of Technology
Harbin, Heilongjiang, China

Dianguo Xu
Harbin Institute of Technology
Harbin, Heilongjiang, China

ISBN 978-981-15-8568-5 ISBN 978-981-15-8566-1 (eBook)
https://doi.org/10.1007/978-981-15-8566-1

© The Editor(s) (if applicable) and The Author(s), under exclusive license to Springer Nature Singapore Pte Ltd. 2021
This work is subject to copyright. All rights are solely and exclusively licensed by the Publisher, whether the whole or part of the material is concerned, specifically the rights of translation, reprinting, reuse of illustrations, recitation, broadcasting, reproduction on microfilms or in any other physical way, and transmission or information storage and retrieval, electronic adaptation, computer software, or by similar or dissimilar methodology now known or hereafter developed.
The use of general descriptive names, registered names, trademarks, service marks, etc. in this publication does not imply, even in the absence of a specific statement, that such names are exempt from the relevant protective laws and regulations and therefore free for general use.
The publisher, the authors and the editors are safe to assume that the advice and information in this book are believed to be true and accurate at the date of publication. Neither the publisher nor the authors or the editors give a warranty, expressed or implied, with respect to the material contained herein or for any errors or omissions that may have been made. The publisher remains neutral with regard to jurisdictional claims in published maps and institutional affiliations.

This Springer imprint is published by the registered company Springer Nature Singapore Pte Ltd.
The registered company address is: 152 Beach Road, #21-01/04 Gateway East, Singapore 189721, Singapore

Preface

Reduced DC-link capacitance motor drive system has been developed for its higher reliability, higher power density, lower cost, and longer lifetime. Compared with the conventional drive system equipped with large volume electrolytic capacitors, the DC-link capacitance is reduced remarkably. The DC-link voltage fluctuates obviously, and the coupling between the grid input side and the inverter output side is enhanced. Many techniques have been developed to improve the drive system performance in the last few decades.

The book focuses on the advanced control of reduced dc-link capacitance AC motor drives. The proposed control strategies are verified by experimental results, which include high power factor control, drive system stability control, beat phenomenon suppression, enhanced flux-weakening control, anti-overvoltage control, etc. The major features of this book are the systematic analysis, effective and optimized control of the practical issues in industry application, which could help the readers to learn the reduced dc-link capacitance PMSM drives and promote the drive system application. This book could benefit researchers, engineers, and students in the field of AC motor drives.

Harbin, China

Gaolin Wang
Nannan Zhao
Guoqiang Zhang
Dianguo Xu

Contents

Nomenclature

DSP	Digital Signal Processor
EEMF	Extended Electromotive Force
EMF	Electromotive Force
FOC	Field-Oriented Control
HPF	High-Pass Filter
IPMSM	Interior Permanent Magnet Synchronous Motor
LPF	Low-Pass Filter
MTPA	Maximum Torque per Ampere
PLL	Phased-Lock Loop
PMSM	Permanent Magnet Synchronous Motor
SPMSM	Surface-Mounted Permanent Magnet Synchronous Motor
SVPWM	Space Vector Pulse Width Modulation
VSI	Voltage Source Inverter
α, β	Stationary Coordinate System
d, q	Rotating Coordinate System
p	Differential Operator D/Dt
s	Laplace Transform Variable
δ	Sliding-Mode Boundary Layer
z	Sliding-Mode Control Function
R_s	Stator Resistance
$\boldsymbol{u}_s$	Stator Voltage Vector
$\boldsymbol{i}_s$	Stator Current Vector
u_d, u_q	d- and q-axis Stator Voltage
i_d, i_q	d- and q-axis Stator Current
e_d, e_q	Extended Back EMF in d- and q-axis
u_α, u_β	α- and β-axis Stator Voltages

i_α, i_β	α- and β-axis Stator Currents
u_a, u_b, u_c	a-, b- and c-axis Stator Phase Voltages
i_a, i_b, i_c	a-, b- and c-axis Stator Phase Currents
e_u, e_v, e_w	Three-phase Grid Voltage
e_g	Single-phase Grid Voltage
u_{dc}	DC-Link Voltage of the Drive System
$u_{dc,0}$	Mean Value of DC-link Voltage
ψ_f	Permanent Flux Linkage
ψ_α, ψ_β	α- and β-axis Stator Flux Components
ψ_d, ψ_q	d- and q-axis Stator Flux Components
L_d, L_q	d- and q-axis Inductances
L_a, L_b, L_c	a-, b- and c-axis Self-Inductances
T_e	Electromagnetic Torque
T_L	Load Torque
θ_e	Rotor Electrical Position
ω_e, ω_r	Rotor Electrical And Mechanical Speeds.
P_n	Pole Pairs Number
J	Inertia Moment
B	Coefficient of Viscous Friction
*	Reference Value
^	Estimated Value
~	Estimation Error

Chapter 1
Basic Knowdge of AC Motor Drives

1.1 Structure and Mathematical Model

1.1.1 PMSM Applications

Permanent magenent synchronous machine (PMSM) has been widely developed in the industry and household applications for its high-power density, high efficiency, high power factor, and excellent control performance. PMSM has been applied in the following areas [1]:

(1) **The aeronautic and astronautic applications**. The requirements of the power density and the control performance are very strict in the aeronautic and astronautic applications. Hence the PMSM is a suitable choice for the aeronautic applications to reduce the launching cost of the whole system.
(2) **The electric vehicle (EV) applications**. EV is the potential traffic solution, which is supported by worldwide governments and companies. Due to the rapid development of EV, PMSM is widely applied as the power supply element due to its high-power density and high efficiency.
(3) **The industry applications**. For high-performance equipments, the excellent control performance of PMSM could satisfy the requirements, such as the computer numerical control (CNC), the robot drives, and the servo systems.
(4) **The domestic appliance**. The high-power density and high efficiency of PMSM could reduce the power cost of the drive system effectively, which is an important concern in the household applications, such as the vacuum cleaner, the washing machine, the air-conditioner, and the refrigerator, etc.

Hence PMSM has a promising future in the energy conversion applications, which deserves to be studied further. However, there are also some disadvantages of PMSM infulencing its applications:

(1) **The cost of the drive system**. The cost of PMSM drive system is relatively high compared with the induction machine (IM) drive system. As for the general

© The Author(s), under exclusive license to Springer Nature Singapore Pte Ltd. 2021

G. Wang et al., *Reduced DC-link Capacitance AC Motor Drives*,
https://doi.org/10.1007/978-981-15-8566-1_1

industry applications, the drive system cost is an important issue, hence it is necessary to optimize the design of the PMSM to reduce the cost.

(2) **Irreversible demagnetization**. Irreversible demagnetization could occur when PMSM operates in excessively high temperature and low temperature, which could reduce the performance of the motor or even disable the normal operation.

1.1.2 PMSM Structure

According to the installation position of the permanent magnet on the rotor, PMSM can be divided into two types: surface mounted PMSM (SPMSM) and interior PMSM (IPMSM). The permanent magnet of SPMSM is mounted on the surface of the rotor core, whereas the permanent magnet of IPMSM is located inside the rotor. For each rotor structure, a sinusoidal distribution of air gap magnetic field is expected.

As shown in Fig. 1.1, the rotor magnetic circuit structure of SPMSM can be divided into the raised type and the plug-in type. The relative permeability of permanent magnet materials is close to 1, which leads to the symmetric orthogonal magnetic path of PMSM, and the impedances of the *d*- and *q*-axis are the same ($X_d = X_q$). SPMSM has the advantages such as simple structure, lower manufacturing cost, and smaller moment of inertia. Besides, the optimal design can be achieved for SPMSM making the air gap magnetic field close to the sinusoidal distribution. Therefore, the performance of the motor and the entire drive system could be improved correspondingly.

Figure 1.2 shows the interior rotor magnetic circuit structure. Permanent magnets in the interior rotor are protected by pole pieces. The reluctance torque generated by the asymmetry of the rotor magnetic circuit structure can improve the overload

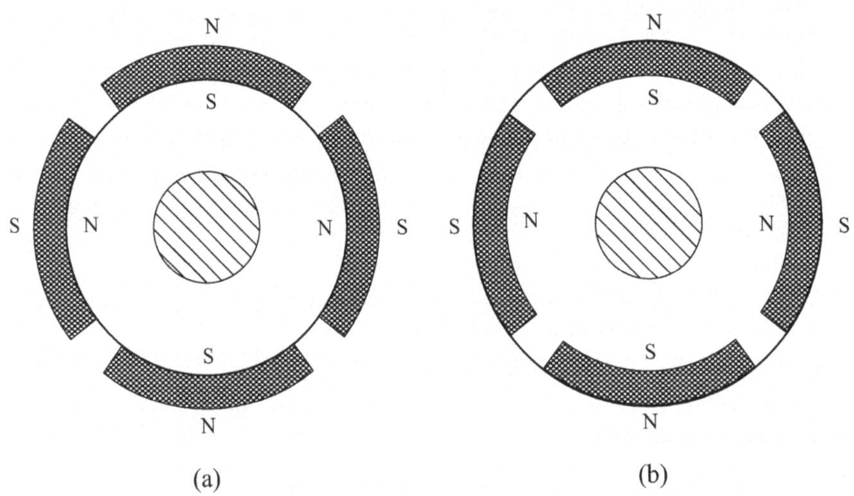

Fig. 1.1 Surface rotor magnetic circuit structure. **a** Raised type. **b** Plug-in type

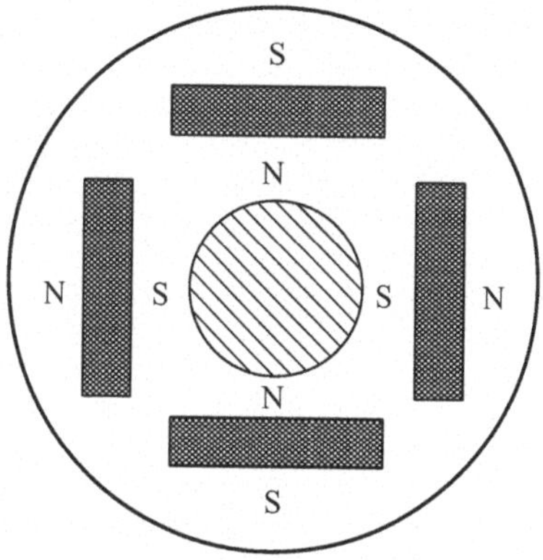

Fig. 1.2 Interior rotor magnetic circuit structure

capacity and the power density of the motor. Meanwhile, the d-axis inductance of IPMSM is usually larger than that of SPMSM, so it is easy to weaken the magnetic field and expand the operation speed. It should be noted that the X_d, X_q and X_d/X_q (salient ratio) of IPMSM are different when the rotor magnetic circuit structure is different. Larger salient ratio could improve the pull-in synchronization ability, the reluctance torque and the overload capability of motor.

1.1.3 PMSM Mathematical Model in Three-Phase Coordinate Frame

The mathematical modeling of PMSM is the key to realize vector control. To simplify the analysis, it is necessary to make the following assumptions:

(1) The three-phase stator windings are distributed symmetrically in space and the discrete properties of its structure are ignored. Therefore, the armature reactive magnetomotive force generated by stator windings in the air gap is sinusoidal, and the induced back electromotive force (EMF) is also the sinusoidal wave.
(2) The internal permeability of a permanent magnet is consistent with that of air, and the EMF produced in the air gap is also a sinusoidal distribution.
(3) The iron loss, the terminal effect and the magnetic saturation effect are ignored. The constant rotor permeability is infinite.
(4) Regardless of the influence of temperature and load effect on the motor parameters, there is no damper winding on the rotor.

Figure 1.3 shows the IPMSM physical model. The voltage equation of its three-phase windings can be presented as

$$\begin{bmatrix} u_a \\ u_b \\ u_c \end{bmatrix} = \begin{bmatrix} R_s & 0 & 0 \\ 0 & R_s & 0 \\ 0 & 0 & R_s \end{bmatrix} \begin{bmatrix} i_a \\ i_b \\ i_c \end{bmatrix} + p \begin{bmatrix} \psi_a \\ \psi_b \\ \psi_c \end{bmatrix} \tag{1.1}$$

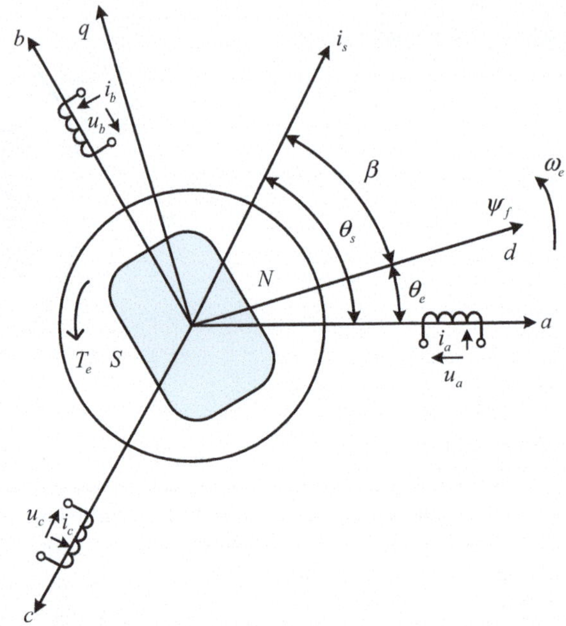

Fig. 1.3 Physical model of PMSM

where u_a, u_b, and u_c are the three-phase voltages, ψ_a, ψ_b and ψ_c are flux linkages of *a-b-c* three-phase windings, R_s is the stator resistance, and p is differential operator *d*/*dt*.

The three-phase flux linkages could be presented as

$$\begin{bmatrix}\psi_a\\ \psi_b\\ \psi_c\end{bmatrix} = \boldsymbol{L}_{abc}\begin{bmatrix}i_a\\ i_b\\ i_c\end{bmatrix} + \psi_f\begin{bmatrix}\cos(\theta_e)\\ \cos(\theta_e - \frac{2\pi}{3})\\ \cos(\theta_e + \frac{2\pi}{3})\end{bmatrix}$$
$$= \begin{bmatrix}L_a & M_{ab} & M_{ac}\\ M_{ba} & L_b & M_{bc}\\ M_{ca} & M_{cb} & L_c\end{bmatrix}\begin{bmatrix}i_a\\ i_b\\ i_c\end{bmatrix} + \begin{bmatrix}\psi_{fa}\\ \psi_{fb}\\ \psi_{fc}\end{bmatrix} \quad (1.2)$$

where ψ_{fa}, ψ_{fb} and ψ_{fc} are the fluxes of the permanent magnet passing through *a-b-c* windings, $\boldsymbol{L}_{abc}$ is the inductance matrix, ψ_f is the flux of the permanent magnet, θ_e is the rotor electrical position. Since the air gap of IPMSM is not uniform, the self-inductances (L_a, L_b, and L_c) and mutual inductances (M_{ab}, M_{ac}, M_{ba}, M_{bc}, M_{ca}, and M_{cb}) of *a-b-c* three-phase windings are periodic functions of the rotor position as shown in (1.3).

$$\begin{cases} L_a = L_{s0} + L_{s2}\cos(2\theta_e)\\ L_b = L_{s0} + L_{s2}\cos 2\left(\theta_e - \frac{2\pi}{3}\right)\\ L_c = L_{s0} + L_{s2}\cos 2\left(\theta_e + \frac{2\pi}{3}\right)\end{cases}$$

$$\begin{cases} M_{ab} = M_{ba} = -M_{s0} + M_{s2}\cos 2\left(\theta_e + \frac{2\pi}{3}\right) \\ M_{bc} = M_{cb} = -M_{s0} + M_{s2}\cos(2\theta_e) \\ M_{ca} = M_{ac} = -M_{s0} + M_{s2}\cos 2\left(\theta_e - \frac{2\pi}{3}\right) \end{cases} \tag{1.3}$$

where L_{s0}, L_{s2} are the average magnitude and the second harmonic of the self-inductance, M_{s0}, and M_{s2} are the average magnitude and the second harmonic of the mutual inductance, respectively.

1.2 Space Vector and Coordinate Transformation

1.2.1 Introduction of PMSM Space Vector

Let $\boldsymbol{\alpha}_1$, $\boldsymbol{\alpha}_2$, ..., $\boldsymbol{\alpha}_n$ be the linear independent vectors of the vector space $\boldsymbol{F}$ [2, 3], where

$$\boldsymbol{\alpha}_1 = \begin{bmatrix} \alpha_{11} \\ \alpha_{12} \\ \vdots \\ \alpha_{1n} \end{bmatrix}, \boldsymbol{\alpha}_2 = \begin{bmatrix} \alpha_{21} \\ \alpha_{22} \\ \vdots \\ \alpha_{2n} \end{bmatrix}, \ldots, \boldsymbol{\alpha}_n = \begin{bmatrix} \alpha_{n1} \\ \alpha_{n2} \\ \vdots \\ \alpha_{nn} \end{bmatrix}. \tag{1.4}$$

Then any vector $\boldsymbol{x}$ can be represented as

$$\mathbf{x} = x_1\boldsymbol{\alpha}_1 + x_2\boldsymbol{\alpha}_2 + \cdots + x_n\boldsymbol{\alpha}_n. \tag{1.5}$$

The vectors $\boldsymbol{\alpha}_1, \boldsymbol{\alpha}_2, \ldots, \boldsymbol{\alpha}_n$ are a basis of $\boldsymbol{F}$ and $(x_1, x_2, \ldots, x_n)^{\mathrm{T}}$ is the coordinate of x, where $\boldsymbol{F}$ is called the n-dimensional vector space. It can be seen that there are many different bases of $\boldsymbol{F}$, whereas the coordinate representation of $\boldsymbol{x}$ is unique in a particular basis. Equation (1.5) could be presented as

$$\mathbf{x} = (\boldsymbol{\alpha}_1, \boldsymbol{\alpha}_2, \ldots \boldsymbol{\alpha}_n)\begin{bmatrix} x_1 \\ x_2 \\ \vdots \\ x_n \end{bmatrix}. \tag{1.6}$$

Equation (1.6) is the general expression in the three-phase PMSM drive system. According to (1.1) and (1.2), the three-phase voltage could be generalized as

$$\begin{bmatrix} u_a \\ u_b \\ u_c \end{bmatrix} = \mathbf{R}\begin{bmatrix} i_a \\ i_b \\ i_c \end{bmatrix} + \mathbf{L}_{abc}\frac{d}{dt}\begin{bmatrix} i_a \\ i_b \\ i_c \end{bmatrix} + \psi_f \mathbf{f}(\theta_e) \tag{1.7}$$

where $\boldsymbol{R}$ and $f(\theta_e)$ are the resistor matrix, and the expression related to the rotor position, respectively.

As for the three-phase SPMSM, the inductance matrix mainly contains the self-inductance and the mutual inductance of the three phase windings. As the air gap of the SPMSM is uniform, the self-inductance and the mutual inductance of the three phase windings are constant values and independent of the rotor position. As for the IPMSM, the air gap is asymmetric. Hence the self-inductance and the mutual inductance of the three phase windings are closely related to the rotor position. The inductance matrices of both the SPMSM and the IPMSM are coupling and dependent of the rotor position.

1.2.2 Coordinate Transformation

The three-phase currents i_a, i_b, and i_c of the three-phase PMSM could be presented as

$$\begin{cases} i_a = I_m \cos \omega_e t \\ i_b = I_m \cos\left(\omega_e t - \frac{2\pi}{3}\right) \\ i_c = I_m \cos\left(\omega_e t - \frac{2\pi}{3}\right) \end{cases} \tag{1.8}$$

where I_m and ω_e are the magnitude of the motor current and the rotor electrical speed, respectively. The symmetric three-phase currents could be presented in the three-dimensional coordinate frame:

$$\begin{aligned} \mathbf{I} &= i_a \mathbf{i}_A + i_b \mathbf{i}_B + i_c \mathbf{i}_C \\ &= I_m \cos(\omega_e t)\mathbf{i}_A + I_m \cos\left(\omega_e t - \frac{2\pi}{3}\right)\mathbf{i}_B + I_m \cos\left(\omega_e t - \frac{2\pi}{3}\right)\mathbf{i}_C \end{aligned} \tag{1.9}$$

where $\boldsymbol{i}_A$, $\boldsymbol{i}_B$, and $\boldsymbol{i}_C$ are the unit vector of the three-dimensional orthogonal frame, respectively. The magnitude of the vector $\boldsymbol{I}$ could be presented as

$$|\mathbf{I}| = \sqrt{i_a^2 + i_b^2 + i_c^2} = I_m \sqrt{\cos^2 \omega_e t + \cos^2 \omega_e t + \cos^2 \omega_e t} = \sqrt{\frac{3}{2}} I_m. \tag{1.10}$$

It can be seen that the vector $\boldsymbol{I}$ rotates around the original point of the three-dimensional orthogonal frame. The magnitude and the angular frequency of $\boldsymbol{I}$ are $\sqrt{3/2} I_m$ and ω_e, respectively. The trajectory of the rotation is a circle and the projection of the trajectory in the frame is the three-phase current, which is shown in Fig. 1.4.

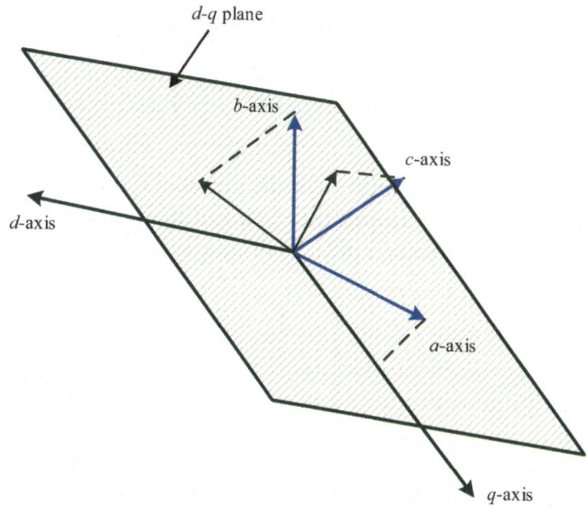

Fig. 1.4 The current vector $\boldsymbol{I}$ in the three-dimensional coordinate frame

The trajectory is located in a plane in the three-dimensional coordinate frame, which is defined as the d-q plane. All the transformation of the coordinate frame is carried out in this plane.

It can be seen that the α-β orthogonal coordinate frame is defined as shown in Fig. 1.5. The coordinate transformation is to transform the vector in the a-b-c frame to the d-q orthogonal coordinate frame. According to the previous analysis, the vector $\boldsymbol{I}$ rotates around the original point with the angular frequency ω_e.

The coordinate transformation could be derived from the geometric view. The relationship between the α-β orthogonal coordinate frame and the a-b-c coordinate frame could be presented as

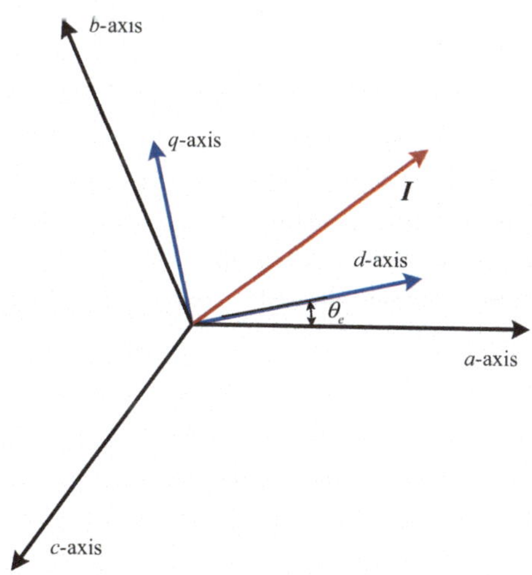

Fig. 1.5 The current vector $\boldsymbol{I}$ in the d-q frame

$$\begin{aligned} N_2 i_\alpha &= N_1 i_a + N_1 i_b \cos\left(\frac{2\pi}{3}\right) + N_1 i_c \cos\left(\frac{4\pi}{3}\right) \\ N_2 i_\beta &= N_1 i_b sin\left(\frac{2\pi}{3}\right) + N_1 i_c sin\left(\frac{4\pi}{3}\right) \end{aligned} \tag{1.11}$$

where i_α, i_β, N_1, and N_2 are the currents in the α- and β- axis, the number of turns in the windings, respectively. Hence the transformation matrix between the *a-b-c* coordinate frame and the α-β orthogonal coordinate frame could be presented as

$$\begin{bmatrix} i_\alpha \\ i_\beta \\ i_0 \end{bmatrix} = \frac{N_1}{N_2} \begin{bmatrix} 1 & -\frac{1}{2} & -\frac{1}{2} \\ 0 & \frac{\sqrt{3}}{2} & -\frac{\sqrt{3}}{2} \\ \frac{1}{\sqrt{2}} & \frac{1}{\sqrt{2}} & \frac{1}{\sqrt{2}} \end{bmatrix} \begin{bmatrix} i_a \\ i_b \\ i_c \end{bmatrix} \tag{1.12}$$

where i_0 is the zero-sequence component of the three-phase current. As for the three-phase star connected PMSM, the sum of the three-phase currents is zero and the corresponding zero-sequence component i_0 is also zero. Hence i_0 can be ignored to make the derivation process clearer. The simplified Clarke transformation could be presented as

$$\begin{bmatrix} i_\alpha \\ i_\beta \end{bmatrix} = \frac{N_1}{N_2} \begin{bmatrix} 1 & -\frac{1}{2} & -\frac{1}{2} \\ 0 & \frac{\sqrt{3}}{2} & -\frac{\sqrt{3}}{2} \end{bmatrix} \begin{bmatrix} i_a \\ i_b \\ i_c \end{bmatrix}. \tag{1.13}$$

Considering the power invariance principle, which means that the drive system power in the *a-b-c* coordinate frame and the α-β coordinate frame remains constant, $\frac{N_1}{N_2}$ could be obtained as $\sqrt{\frac{2}{3}}$. And the Clarke transformation could be presented as

$$\begin{bmatrix} i_\alpha \\ i_\beta \\ i_0 \end{bmatrix} = \sqrt{\frac{2}{3}} \begin{bmatrix} 1 & -\frac{1}{2} & -\frac{1}{2} \\ 0 & \frac{\sqrt{3}}{2} & -\frac{\sqrt{3}}{2} \\ \frac{1}{\sqrt{2}} & \frac{1}{\sqrt{2}} & \frac{1}{\sqrt{2}} \end{bmatrix} \begin{bmatrix} i_a \\ i_b \\ i_c \end{bmatrix}. \tag{1.14}$$

If considering the phase magnitude invariance method, the magnitude of the voltages and currents in the *a-b-c* coordinate frame and the α-β coordinate frame remains constant, whereas the torque and the power should be multiplied by 3/2 in the α-β frame to obtain the same values in the *a-b-c* coordinate frame. The transformation matrix in the α-β frame could be presented as

$$\begin{bmatrix} i_\alpha \\ i_\beta \\ i_0 \end{bmatrix} = \frac{2}{3} \begin{bmatrix} 1 & -\frac{1}{2} & -\frac{1}{2} \\ 0 & \frac{\sqrt{3}}{2} & -\frac{\sqrt{3}}{2} \\ \frac{1}{\sqrt{2}} & \frac{1}{\sqrt{2}} & \frac{1}{\sqrt{2}} \end{bmatrix} \begin{bmatrix} i_a \\ i_b \\ i_c \end{bmatrix}. \tag{1.15}$$

The α-β coordinate frame is the stationary frame, and the elements of the inductance matrix are closely related to the rotor position, hence it is essential to eliminate the change caused by different values of the rotor position. The Park transformation is applied to transform the variables in the stationary α-β coordinate frame to the rotating d-q coordinate frame. The transformation can be presented as

$$\begin{bmatrix} i_d \\ i_q \end{bmatrix} = \begin{bmatrix} \cos\theta_e & -\sin\theta_e \\ \sin\theta_e & \cos\theta_e \end{bmatrix} \begin{bmatrix} i_\alpha \\ i_\beta \end{bmatrix}$$
$$\begin{bmatrix} i_\alpha \\ i_\beta \end{bmatrix} = \begin{bmatrix} \cos\theta_e & \sin\theta_e \\ -\sin\theta_e & \cos\theta_e \end{bmatrix} \begin{bmatrix} i_d \\ i_q \end{bmatrix}. \tag{1.16}$$

where i_d and i_q are the motor currents in the d-q coordinate frame.

1.2.3 PMSM Mathematical Model in Different Coordinate Frames

As can be seen from (1.3), since the inductances of the a-b-c three-phase windings are the periodic functions of the rotor position, the voltage equations are the time-varying differential forms. It is essential to apply the decoupling control method to transform the voltage equations.

By applying the Clarke transformation, (1.1) and (1.2) are transformed into the α-β stationary coordinate system. The voltage equation, the flux linkage and the inductance matrix of α-β axes could be presented as

$$\begin{bmatrix} u_\alpha \\ u_\beta \end{bmatrix} = R_s \begin{bmatrix} i_\alpha \\ i_\beta \end{bmatrix} + p \begin{bmatrix} \psi_\alpha \\ \psi_\beta \end{bmatrix} \tag{1.17}$$

$$\begin{bmatrix} \psi_\alpha \\ \psi_\beta \end{bmatrix} = L_{\alpha\beta} \begin{bmatrix} i_\alpha \\ i_\beta \end{bmatrix} + \psi_f \begin{bmatrix} \cos\theta_e \\ \sin\theta_e \end{bmatrix} \tag{1.18}$$

$$\mathbf{L}_{\alpha\beta} = \begin{bmatrix} L_\alpha & M_{\alpha\beta} \\ M_{\beta\alpha} & L_\beta \end{bmatrix} = \begin{bmatrix} \Sigma L + \Delta L\cos(2\theta_e) & \Delta L\sin(2\theta_e) \\ \Delta L\sin(2\theta_e) & \Sigma L - \Delta L\cos(2\theta_e) \end{bmatrix} \tag{1.19}$$

where u_α and u_β are the stator voltage components of α-β axes, ψ_α and ψ_β are the stator flux components of α-β axes, $\mathbf{L}_{\alpha\beta}$ is the inductance matrix of α-β axes, $\Sigma L = (L_d + L_q)/2$ and $\Delta L = (L_d - L_q)/2$, L_d and L_q are the equivalent inductances of the d-q axis.

As can be seen from (1.19), the inductances at α-β axes are still periodic functions of the rotor position. By applying the Park transformation, (1.17) and (1.18) can be transformed into the d-q coordinate frame. The voltage equation, the flux linkage and the inductance matrix of the d-q coordinate frame could be presented as

$$\begin{bmatrix} u_d \\ u_q \end{bmatrix} = R_s \begin{bmatrix} i_d \\ i_q \end{bmatrix} + p \begin{bmatrix} \psi_\alpha \\ \psi_\beta \end{bmatrix} + \omega_e \begin{bmatrix} -\psi_q \\ \psi_d \end{bmatrix} \tag{1.20}$$

$$\begin{bmatrix} \psi_d \\ \psi_q \end{bmatrix} = L_{dq} \begin{bmatrix} i_d \\ i_q \end{bmatrix} + \begin{bmatrix} \psi_f \\ 0 \end{bmatrix} \tag{1.21}$$

$$\mathbf{L}_{dq} = \begin{bmatrix} L_d & 0 \\ 0 & L_q \end{bmatrix} \tag{1.22}$$

where u_d and u_q are the stator voltage components of d-q axis, ψ_d and ψ_q are the stator flux components of d-q axis, L_d, and L_q is the inductance of d-q axis. The model of SPMSM is also represented by (1.20)–(1.22) with $L_d = L_q$.

As can be seen from (1.22), the inductance matrix in the d-q axis becomes the constant matrix and is no longer the periodic function of the rotor position, thus the decoupling of the d-q axis could be achieved. The phasor diagram of three-phase PMSM is shown in Fig. 1.6, the stator voltage phasor $\dot{U}_s$ could be presented as

$$\dot{U}_s = R_s \dot{I}_s + j\omega_e L_d \dot{I}_d + j\omega_e L_q \dot{I}_q + \dot{E}_0 \tag{1.23}$$

where $\dot{I}_d$, $\dot{I}_q$ and $\dot{E}_0$ are the d-q axis current phasor and the EMF phasor, the stator current phasor $\dot{I}_s$ is the sum of $\dot{I}_d$ and $\dot{I}_q$.

The electromagnetic torque T_e in d-q axis of PMSM can be expressed as

$$T_e = \frac{3}{2} P_n \left[\psi_f i_q + (L_d - L_q) i_d i_q \right] \tag{1.24}$$

where P_n is the pole pairs number. The electromechanical dynamic characteristics could be presented as

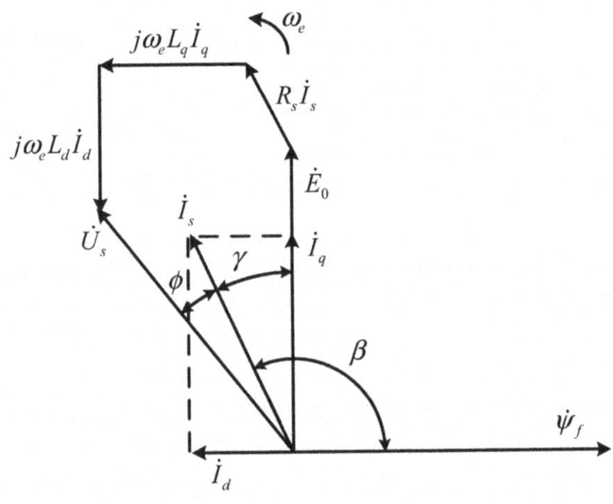

Fig. 1.6 Phasor diagram of three-phase PMSM

$$T_e - T_L = J\frac{d\omega_r}{dt} + B\omega_r \tag{1.25}$$

where T_L is the load torque, J is the inertia moment, B is the coefficient of viscous friction, and $\omega_r = \omega_e/P_n$ is rotor mechanical speed.

1.3 Space Vector Pulse Width Modulation (SVPWM)

1.3.1 Principle and Realization of SVPWM

SVPWM is often used in AC motor drives. The main purpose of classical SVPWM is to make the output fundamental voltage of the inverter close to the sinusoidal wave. SVPWM technology uses the different combinations of control signals of each bridge arm in the inverter to make the fundamental voltage vector trajectory as close to a circle as possible. The equivalent switch model is shown in Fig. 1.7.

Usually, the 180° conduction mode is adopted in voltage source inverters (VSIs). S_A, S_B and S_C are used to represent the states of the three arms, respectively. State 1 indicates that the upper bridge arm is turned on, and state 0 indicates that the lower bridge arm is turned on. Eight switching modes can be obtained with the changes of the three arms and each switching mode corresponds to a basic voltage vector. There are totally eight basic voltage vectors formed, including six non-zero vectors and two zero vectors.

In order to understand the principle of SVPWM, one switching mode is selected to describe in detail. When transistors 1, 2, 3 are in the open state, which means $S_A = S_B = 1$, and $S_C = 0$. Therefore, the equivalent circuit of the inverter can be represented as shown in Fig. 1.8.

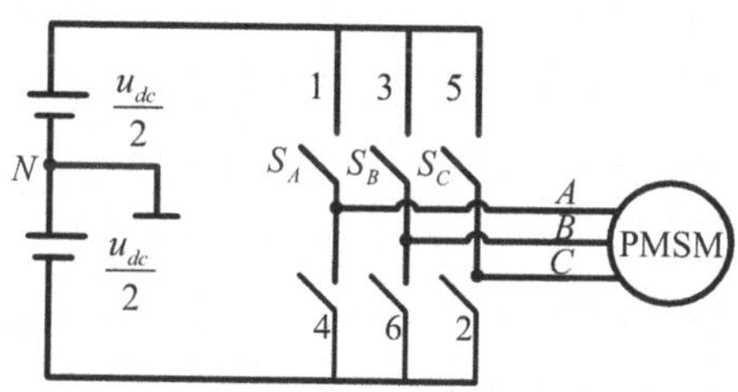

Fig. 1.7 PWM equivalent switch model

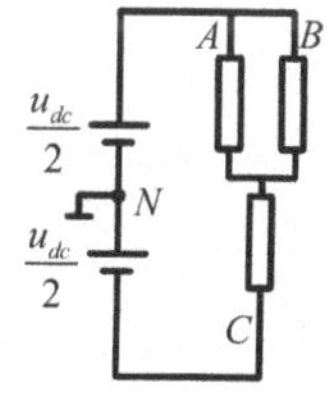

Fig. 1.8 Equivalent circuit diagram of inverter when $S_A = S_B = 1$, and $S_C = 0$

Under this circumstance, the voltage drop on each phase can be denoted as

$$u_{AN} = \frac{1}{3}u_{dc}, u_{BN} = \frac{1}{3}u_{dc}, u_{CN} = -\frac{2}{3}u_{dc}. \tag{1.26}$$

The voltage vector $\boldsymbol{u}$ synthetized by $\boldsymbol{u}_{AN}$, $\boldsymbol{u}_{BN}$, and $\boldsymbol{u}_{CN}$ is shown as in Fig. 1.9. In the α-β axes, the voltage vector $\boldsymbol{u}$ can be expressed as

$$\boldsymbol{u} = k\left(\boldsymbol{u}_{AN} + \boldsymbol{u}_{BN}e^{j\frac{2}{3}\pi} + \boldsymbol{u}_{CN}e^{j\frac{4}{3}\pi}\right) \tag{1.27}$$

where k is the transformation coefficient to transform the vector from the three-phase stationary coordinate system to the two-phase stationary coordinate system. The transformation can be divided into equal power transformation and equal quantity transformation. Here, the equal quantity transformation is used, and $k = 2/3$.

$$\boldsymbol{u} = \frac{2}{3}(\boldsymbol{u}_{AN} + \boldsymbol{u}_{BN}e^{j\frac{2}{3}\pi} + \boldsymbol{u}_{CN}e^{j\frac{4}{3}\pi}) \tag{1.28}$$

Substituting (1.26) into (1.27), the voltage vector can be expressed as

$$\boldsymbol{u} = \frac{2}{3}u_{dc}e^{j\frac{1}{3}\pi}. \tag{1.29}$$

The corresponding voltage vectors of each switching state can be obtained through the same method as shown in Table 1.1. As can be seen in Fig. 1.10, the six non-zero voltage vectors and the two zero voltage vectors form a voltage hexagon, and the amplitude of each voltage vector is $2u_{dc}/3$. Meanwhile, the voltage hexagon is divided into six sectors by the voltage vectors.

If only eight basic voltage vectors are used individually, a hexagonal rotating voltage vector can be obtained. This is far away from the circular rotating fundamental voltage vector as expected. The additional intermediate voltage vectors should be formed to achieve the circular fundamental voltage vector trajectory. Actually, the

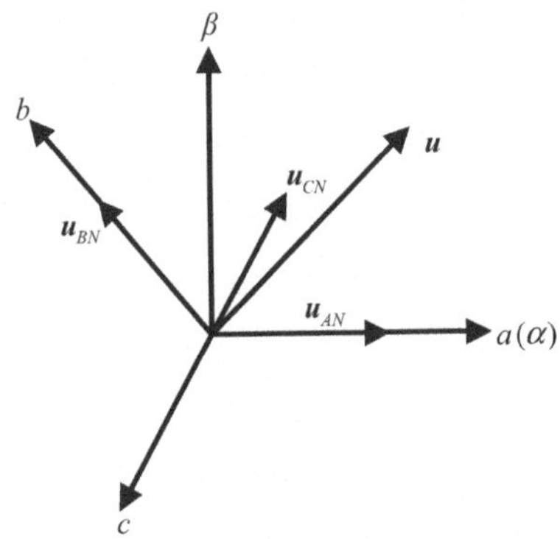

Fig. 1.9 Diagram of voltage vector when $S_A = S_B = 1$, $S_C = 0$

Table 1.1 Basic voltage vectors of SVPWM modulation strategy

$S_AS_BS_C$	A phase voltage	B phase voltage	C phase voltage	Vector expression	Vector symbol
000	0	0	0	0	u_0
001	$-\frac{1}{3}u_{dc}$	$-\frac{1}{3}u_{dc}$	$\frac{2}{3}u_{dc}$	$\frac{2}{3}u_{dc}e^{j\frac{4}{3}\pi}$	u_1
010	$-\frac{1}{3}u_{dc}$	$\frac{2}{3}u_{dc}$	$-\frac{1}{3}u_{dc}$	$\frac{2}{3}u_{dc}e^{j\frac{2}{3}\pi}$	u_2
011	$-\frac{2}{3}u_{dc}$	$\frac{1}{3}u_{dc}$	$\frac{1}{3}u_{dc}$	$\frac{2}{3}u_{dc}e^{j\pi}$	u_3
100	$\frac{2}{3}u_{dc}$	$-\frac{1}{3}u_{dc}$	$-\frac{1}{3}u_{dc}$	$\frac{2}{3}u_{dc}e^{j0}$	u_4
101	$\frac{1}{3}u_{dc}$	$-\frac{2}{3}u_{dc}$	$\frac{1}{3}u_{dc}$	$\frac{2}{3}u_{dc}e^{j\frac{5}{3}\pi}$	u_5
110	$\frac{1}{3}u_{dc}$	$\frac{1}{3}u_{dc}$	$-\frac{2}{3}u_{dc}$	$\frac{2}{3}u_{dc}e^{j\frac{1}{3}\pi}$	u_6
111	0	0	0	0	u_7

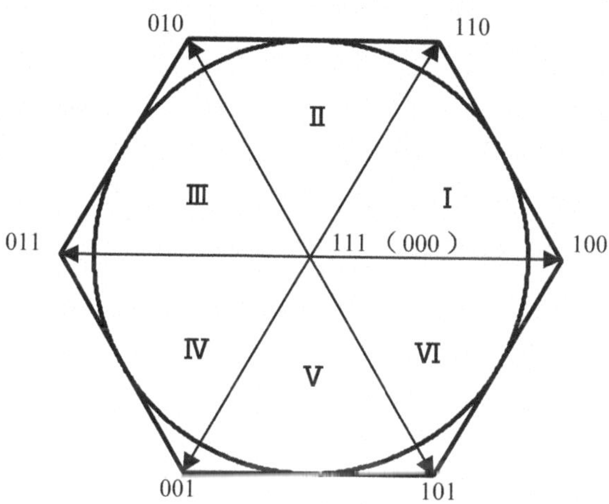

Fig. 1.10 Voltage hexagon and sectors

intermediate vectors can be synthesized by six non-zero voltage vectors and two zero voltage vectors.

It is impossible to realize two switching states at the same time. However, if the switching frequency of the inverter is much higher than the frequency of its output voltage, the desired voltage vector can be synthesized by using the basic voltage vector based on the state-space averaging method. As shown in Fig. 1.11, where $\boldsymbol{u}_x$ and $\boldsymbol{u}_y$ are the two adjacent basic voltage vectors, T_s is the switching time, T_x and T_y are the corresponding opening times. In one switching period, the basic voltage vectors $\boldsymbol{u}_x$ and $\boldsymbol{u}_y$ act for T_x and T_y, respectively. Then the desired voltage vector can be synthesized by the two vectors. Ultimately, the circular voltage vector trajectories can be obtained.

Fig. 1.11 Diagram of space vector constituted by adjacent basic voltage vectors

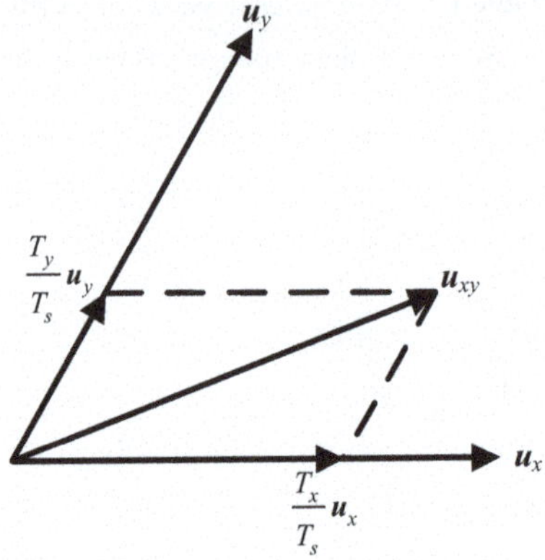

1.3.2 *Evaluation of Maximum Voltage Vector in SVPWM*

In the previous section, the circular fundamental voltage vector trajectory is determined, which means the voltage vector instruction cannot exceed the inscribed circle of the hexagon. However, in some conditions, the utilization rate of the DC-link voltage is desired to be increased. As shown in Fig. 1.12, the voltage hexagon is divided into the linear modulation region and the overmodulation region. In order to expand the output voltage of the inverter, some overmodulation strategies were designed in past decades.

The modulation index M for Pulse-Width Modulation (PWM) inverter is defined here as

$$M = \frac{|\boldsymbol{u}|}{2u_{dc}/\pi}. \tag{1.30}$$

Fig. 1.12 Region division of SVPWM

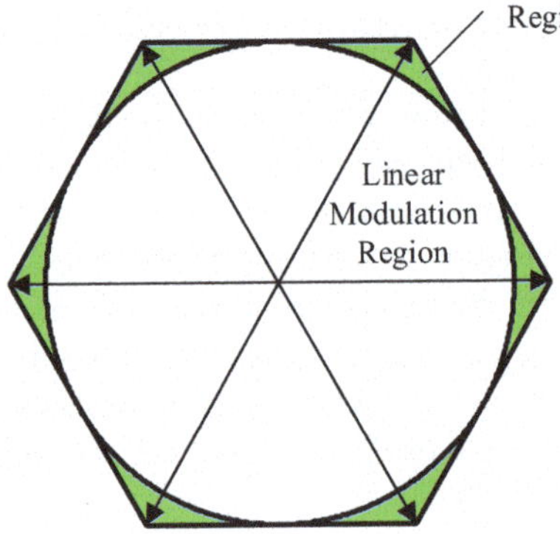

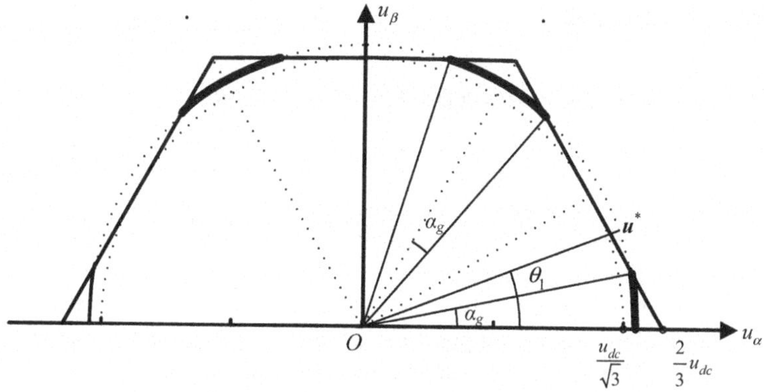

Fig. 1.13 Diagram of overmodulation strategy proposed by Bolognani

When the modulation index exceeds 0.9069, the inverter operates in the overmodulation region. The maximum modulation index is 1.

The basic idea of an overmodulatin scheme is to keep the output voltage vector fixed when the reference voltage vector exceeds the hexagonal boundary [4]. When the reference voltage vector trajectory is within the hexagon, the linear modulation method is used. Since the modulation methods in each sector are similar, sector 1 is discussed here. As shown in Fig. 1.13, θ_1 is the phase of the reference voltage vector $\boldsymbol{u}^*$. The linear modulation method is used before $\boldsymbol{u}^*$ runs to the holding angle α_g. When θ_1 runs from α_g to π/6, the phase of $\boldsymbol{u}^*$ is limited to $\theta = \alpha_g$. And when θ_1 runs from π/6 to π/3-α_g, the phase of $\boldsymbol{u}^*$ is limited to $\theta =$ π/3-α_g. When θ_1 runs from π/3-α_g to π/3, a linear modulation is used again.

The phase θ linked to the phase θ_1 can be expressed as

$$\theta=\begin{cases}\theta_1 & 0 \le \theta_1 \le \alpha_g \\ \alpha_g & \alpha_g \le \theta_1 \le \pi/6 \\ \pi/3-\alpha_g & \pi/6 \le \theta_1 \le \pi/3-\alpha_g \\ \theta_1 & \pi/3-\alpha_g \le \theta_1 \le \pi/3\end{cases}. \tag{1.31}$$

The holding angle α_g can be derived as

$$\alpha_g=\frac{\pi}{6}-\arccos\left(\frac{u_{dc}}{\sqrt{3}|\boldsymbol{u}^*|}\right). \tag{1.32}$$

The idea of this control method is clear. The output voltage vector is limited to the hexagon and can achieve six-step operation when $|\boldsymbol{u}^*|$ reaches $2/3u_{dc}$. The DC-link voltage of the inverter can be effectively utilized. However, the phase error is bigger than the minimum phase error overmodulation method and the minimum amplitude error overmodulation method [5].

1.4 Vector Control

1.4.1 Basic Structure of Vector Control System

The magnetic field of the AC motor must be well managed when it is controlled by closed-loop type. At present, the mainstream high-performance control technology of PMSM drives is Field Orientation Control (FOC). FOC conducts high-performance closed-loop control based on the torque generation of the motor. Its schematic diagram is shown in Fig. 1.14. The core of the control technology is to control the excitation current component i_d and the torque current component i_q of the stator separately in the two-phase rotating coordinate frame.

1.4.2 Principle of Field Orientation Control

Speed & current closed loop is the mainstream of AC motor vector control. The speed loop is used as the outer loop to control the speed of the motor, and its output serves as the reference of the current regulator. The current loop is used to control the torque current and the excitation current of the motor. The controller is usually designed as a classical PID controller, and parameter setting can be realized by engineering optimum method. The block diagram of the control system is shown in Fig. 1.15. It can be seen that the sensorless control is adopted here to obtain the position and speed information instead of using the mechanical position sensors. In this way, the reliability and robustness could be improved and the total cost could be reduced. The model based sensorless control using the electromotive force (EMF) will be introduced in the following.

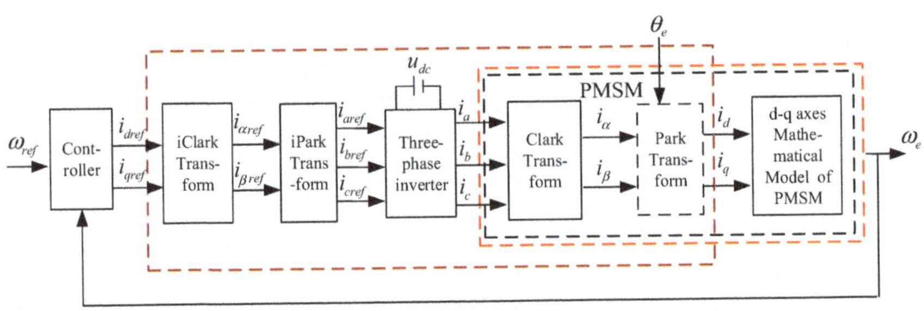

Fig. 1.14 Schematic diagram of vector control system

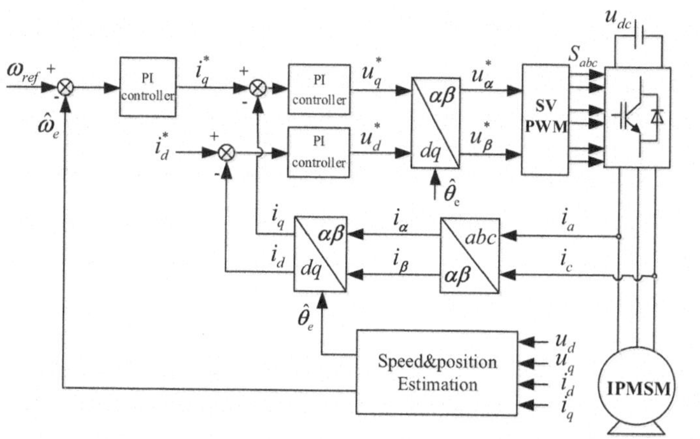

Fig. 1.15 Block diagram of position sensorless vector control system based on EMF observer

1.5 Model Based Sensorless Control

1.5.1 Concept of Extended Electromotive Force

In the synchronous reference frame, the mathematical model of IPMSM can be expressed as (1.17). By transforming this equation into the α-β stationary reference frame, the mathematical model in the stationary coordinate system can be obtained as,

$$\begin{bmatrix} u_\alpha \\ u_\beta \end{bmatrix} = \begin{bmatrix} R_s + pL_\alpha & pM_{\alpha\beta} \\ pM_{\beta\alpha} & R_s + pL_\beta \end{bmatrix} \begin{bmatrix} i_\alpha \\ i_\beta \end{bmatrix} + \omega_e \psi_f \begin{bmatrix} -\sin\theta_e \\ \cos\theta_e \end{bmatrix} \tag{1.33}$$

According to (1.33), the rotor position information of the IPMSM is contained not only in the back-EMF term but also in the inductance matrix due to the rotor saliency. In addition, the inductance matrix contains θ_e term and $2\theta_e$ term, which increases the difficulty of rotor position detection. Therefore, it is not possible for IPMSMs to directly use the voltage equation of the α-β axes for position estimation like SPMSMs.

In order to decouple the rotor position information of IPMSM from the voltage equation in the α-β axes, (1.33) is rewritten to make the coefficient matrix symmetrical. Then the following equivalent transformation is performed:

$$\begin{bmatrix} u_d \\ u_q \end{bmatrix} = \begin{bmatrix} R_s + pL_d & -\omega_e L_q \\ \omega_e L_q & R_s + pL_d \end{bmatrix} \begin{bmatrix} i_d \\ i_q \end{bmatrix} + \begin{bmatrix} 0 \\ (L_d - L_q)(\omega_e i_d - \dot{i}_q) + \omega_e \psi_f \end{bmatrix} \tag{1.34}$$

Define E_{ex} as Extended EMF (EEMF) [6]:

$$E_{ex} = (L_d - L_q)(\omega_e i_d - \dot{i}_q)\omega_e\psi_f. \tag{1.35}$$

Thus, the EEMF E_{ex} includes not only the traditional back EMF term, but also the term generated by polarity saliency of the IPMSM. The concept of EEMF can be extended to the entire synchronous motor category. When $L_d = L_q$, it is the back EMF model of SPMSM. When $\psi_f = 0$, it is the reluctance motor EMF model.

1.5.2 Sliding-Mode Observer Construction

In general, the mathematical model in α-β axes system is usually used to construct the PMSM position observer since it is easy for implementation. Thus, (1.34) is transformed into the α-β stationary coordinate system to obtain the voltage equations based on the EEMF model,

$$\begin{bmatrix} u_\alpha \\ u_\beta \end{bmatrix} = \begin{bmatrix} R_s + pL_d & \omega_e(L_d - L_q) \\ -\omega_e(L_d - L_q) & R_s + pL_d \end{bmatrix} \begin{bmatrix} i_\alpha \\ i_\beta \end{bmatrix} + \begin{bmatrix} e_\alpha \\ e_\beta \end{bmatrix} \tag{1.36}$$

where $e_\alpha = -E_{ex}\sin\theta_e$ and $e_\beta = E_{ex}\cos\theta_e$ are components of EEMF in the α-β axes.

According to (1.36), the rotor position information can be decoupled from the inductance matrix by means of the equivalent transformation and the introduction of the EEMF concept. In this way, the EEMF is the only term that contains the rotor position information. Then the phase information of the EEMF can be used to realize the rotor position observation.

Rewrite the IPMSM voltage Eq. (1.36) as a state equation using the stator currents as state variables,

$$\begin{bmatrix} \dot{i}_\alpha \\ \dot{i}_\beta \end{bmatrix} = \frac{1}{L_d} \begin{bmatrix} -R_s & -\omega_e(L_d - L_q) \\ \omega_e(L_d - L_q) & -R_s \end{bmatrix} \begin{bmatrix} i_\alpha \\ i_\beta \end{bmatrix} + \frac{1}{L_d} \begin{bmatrix} u_\alpha - e_\alpha \\ u_\beta - e_\beta \end{bmatrix}. \tag{1.37}$$

Since the stator current is the only physical quantity that can be directly measured, the stator current error is selected as the sliding surface,

$$s(x) = \begin{bmatrix} \tilde{i}_\alpha \\ \tilde{i}_\beta \end{bmatrix} = \begin{bmatrix} \hat{i}_\alpha - i_\alpha \\ \hat{i}_\beta - i_\beta \end{bmatrix} = 0 \tag{1.38}$$

where the superscript "^" indicates estimated value and "~" indicates the error of observation, which refers to the difference between the observed value and the actual one.

The traditional second order sliding mode observer is constructed. The mathematical model is shown in (1.39). And the block diagram is shown in Fig. 1.16.

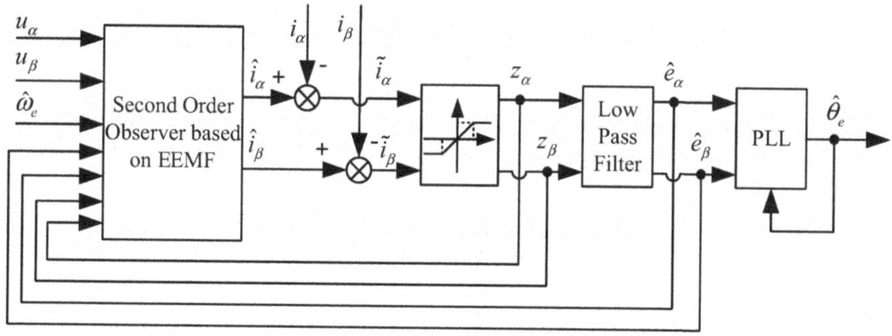

Fig. 1.16 Block diagram of traditional second-order sliding mode observer

$$\begin{bmatrix} \dot{\hat{i}}_\alpha \\ \dot{\hat{i}}_\beta \end{bmatrix} = \frac{1}{L_d}\begin{bmatrix} -R_s & -\hat{\omega}_e(L_d - L_q) \\ \hat{\omega}_e(L_d - L_q) & -R_s \end{bmatrix}\begin{bmatrix} \hat{i}_\alpha \\ \hat{i}_\beta \end{bmatrix} + \frac{1}{L_d}\begin{bmatrix} u_\alpha - \hat{e}_\alpha - z_\alpha \\ u_\beta - \hat{e}_\beta - z_\beta \end{bmatrix} \tag{1.39}$$

where superscripts "^" indicates the observered value, z_α and z_β are sliding mode feedback components:

$$\begin{bmatrix} z_\alpha \\ z_\beta \end{bmatrix} = k\begin{bmatrix} \mathrm{sgn}\left(\hat{i}_\alpha - i_\alpha\right) \\ \mathrm{sgn}\left(\hat{i}_\beta - i_\beta\right) \end{bmatrix} \tag{1.40}$$

where k is the sliding mode gain designed by *Lyapunov* stability analysis.

The observered value of EEMF in α-β axes $(\hat{e}_\alpha, \hat{e}_\beta)$ can be obtained by a low-pass filter (LPF) from the obtained discontinuous switching signals z_α and z_β,

$$\begin{bmatrix} \hat{e}_\alpha \\ \hat{e}_\beta \end{bmatrix} = \frac{\omega_{cl}}{s + \omega_{cl}}\begin{bmatrix} z_\alpha \\ z_\beta \end{bmatrix} \tag{1.41}$$

where ω_{cl} is the cut-off frequency of the LPF, which is usually selected according to the fundamental frequency of the stator current to ensure that the fundamental component is not affected while filtering out the high-frquency components.

1.5.3 Full-Order Sliding-Mode Observer

Normally, the mechanical time constant of the system is much larger than its electromagnetic time constant. In a PWM control period, the angular velocity can be considered as a constant, i.e. $\dot{\omega}_e = 0$. Under this circumstance, the EEMF and its derivative satisfy the following relationship,

$$\begin{bmatrix} \dot{e}_\alpha \\ \dot{e}_\beta \end{bmatrix} = \omega_e \begin{bmatrix} -e_\beta \\ e_\alpha \end{bmatrix}. \tag{1.42}$$

The full-order state equation of IPMSM using the stator currents and the extended back EMFs as state variables can be derived,

$$\begin{bmatrix} \dot{\mathbf{i}} \\ \dot{\mathbf{e}} \end{bmatrix} = \begin{bmatrix} \mathbf{A}_{11} & \mathbf{A}_{12} \\ \mathbf{0} & \mathbf{A}_{22} \end{bmatrix} \begin{bmatrix} \mathbf{i} \\ \mathbf{e} \end{bmatrix} + \begin{bmatrix} \mathbf{B}_1 \\ \mathbf{0} \end{bmatrix} \mathbf{u} \tag{1.43}$$

where

$$\begin{aligned} &\mathbf{i} = \begin{bmatrix} i_\alpha \; i_\beta \end{bmatrix}^{\mathrm{T}},\ \mathbf{e} = \begin{bmatrix} e_\alpha \; e_\beta \end{bmatrix}^{\mathrm{T}},\ \mathbf{u} = \begin{bmatrix} u_\alpha \; u_\beta \end{bmatrix}^{\mathrm{T}} \\ &\mathbf{A}_{11} = -R/L_d \cdot \mathbf{I} + \omega_e (L_d - L_q)/L_d \cdot \mathbf{J} \\ &\mathbf{A}_{12} = -1/L_d \cdot \mathbf{I},\ \mathbf{A}_{22} = \omega_e \cdot \mathbf{J} \\ &\mathbf{B}_1 = 1/L_d \cdot \mathbf{I}, \end{aligned}$$

$$\mathbf{I} = \begin{bmatrix} 1 & 0 \\ 0 & 1 \end{bmatrix},\ \mathbf{J} = \begin{bmatrix} 0 & -1 \\ 1 & 0 \end{bmatrix}.$$

According to the full-order state Eq. (1.43), the full-order sliding mode observer can be established as shown in (1.44). Figure 1.17 shows the block diagram of the full-order sliding mode observer based on EEMF model.

$$\begin{bmatrix} \dot{\hat{\mathbf{i}}} \\ \dot{\hat{\mathbf{e}}} \end{bmatrix} = \begin{bmatrix} \hat{\mathbf{A}}_{11} & \mathbf{A}_{12} \\ \mathbf{0} & \hat{\mathbf{A}}_{22} \end{bmatrix} \begin{bmatrix} \hat{\mathbf{i}} \\ \hat{\mathbf{e}} \end{bmatrix} + \begin{bmatrix} \mathbf{B}_1 \\ \mathbf{0} \end{bmatrix} \mathbf{u} - \frac{1}{L_d} \mathbf{K}\mathrm{sgn}(\mathbf{s}). \tag{1.44}$$

where

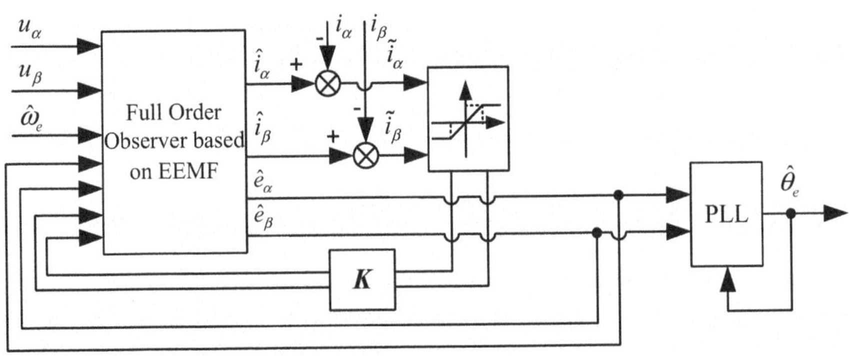

Fig. 1.17 Block diagram of full-order sliding mode observer

$$\begin{aligned}\hat{\mathbf{i}} &= \begin{bmatrix}\hat{i}_\alpha & \hat{i}_\beta\end{bmatrix}^{\mathrm{T}}\\ \hat{\mathbf{e}} &= \begin{bmatrix}\hat{e}_\alpha & \hat{e}_\beta\end{bmatrix}^{\mathrm{T}}\\ \mathbf{s} &= \hat{\mathbf{i}} - \mathbf{i}\end{aligned}$$

$$\begin{aligned}\hat{\mathbf{A}}_{11} &= -R/L_d \cdot \mathbf{I} + \hat{\omega}_e(L_d - L_q)/L_d \cdot \mathbf{J}\\ \hat{\mathbf{A}}_{22} &= \hat{\omega}_e \cdot \mathbf{J}\end{aligned}$$

$$\mathbf{K} = \begin{bmatrix} k & 0\\ 0 & k\\ -m & 0\\ 0 & -m\end{bmatrix}$$

where $\boldsymbol{K}$ is the feedback gain matrix, k and m are the sliding mode observer gains that can be obtained from *Lyapunov* stability analysis.

1.5.4 Stability Analysis of Sliding-Mode Observer

A dynamic model of the current observation error can be obtained by subtracting (1.39) from (1.37):

$$\begin{aligned}\dot{\tilde{i}}_\alpha &= -\frac{R_s}{L_d}\tilde{i}_\alpha - \frac{L_d - L_q}{L_d}\left(\hat{\omega}_e\hat{i}_\beta - \omega_e i_\beta\right) - \frac{1}{L_d}\tilde{e}_\alpha - \frac{1}{L_d}k\mathrm{sgn}\left(\hat{i}_\alpha - i_\alpha\right)\\ \dot{\tilde{i}}_\beta &= -\frac{R_s}{L_d}\tilde{i}_\beta + \frac{L_d - L_q}{L_d}\left(\hat{\omega}_e\hat{i}_\alpha - \omega_e i_\alpha\right) - \frac{1}{L_d}\tilde{e}_\beta - \frac{1}{L_d}k\mathrm{sgn}\left(\hat{i}_\beta - i_\beta\right)\end{aligned} \tag{1.45}$$

where $\tilde{e}_\alpha = \hat{e}_\alpha - e_\alpha$ and $\tilde{e}_\beta = \hat{e}_\beta - e_\beta$ are the α-β components of EEMF.

Normally, when the stator currents reach the sliding mode surface, the observed speed can converge to the actual speed, i.e. $\hat{\omega}_\mathrm{e} = \omega_\mathrm{e}$. Therefore, (1.45) can be further simplified to

$$\begin{aligned}\dot{\tilde{i}}_\alpha &= -\frac{R_s}{L_d}\tilde{i}_\alpha - \frac{L_d - L_q}{L_d}\omega_e\tilde{i}_\beta - \frac{1}{L_d}\tilde{e}_\alpha - \frac{1}{L_d}k\mathrm{sgn}\left(\hat{i}_\alpha - i_\alpha\right)\\ \dot{\tilde{i}}_\beta &= -\frac{R_s}{L_d}\tilde{i}_\beta + \frac{L_d - L_q}{L_d}\omega_e\tilde{i}_\alpha - \frac{1}{L_d}\tilde{e}_\beta - \frac{1}{L_d}k\mathrm{sgn}\left(\hat{i}_\beta - i_\beta\right).\end{aligned} \tag{1.46}$$

In order to analyze the stability of the second-order sliding mode observer, define the positive definite Lyapunov function as:

$$V = \frac{1}{2}s^T s = \frac{1}{2}\left(\tilde{i}_\alpha^2 + \tilde{i}_\beta^2\right). \tag{1.47}$$

In order to enhance the robustness of the observer against disturbance and parameter variation, and to ensure the convergence, it can be known from the conditionality of the sliding mode variable structure that $\dot{V}$ must be negative, i.e. $\dot{V} < 0$,

$$\begin{aligned}\dot{V} =& \mathbf{s}^{\mathrm{T}}\dot{\mathbf{s}}\\ =&\tilde{i}_\alpha \cdot \dot{\tilde{i}}_\alpha + \tilde{i}_\beta \cdot \dot{\tilde{i}}_\beta\\ =&\tilde{i}_\alpha\left\{-\frac{R_s}{L_d}\tilde{i}_\alpha - \frac{L_d - L_q}{L_d}\hat{\omega}_e\tilde{i}_\beta - \frac{1}{L_d}\tilde{e}_\alpha - \frac{1}{L_d}k\mathrm{sgn}\left(\tilde{i}_\alpha\right)\right\}\\ &+\tilde{i}_\beta\left\{-\frac{R_s}{L_d}\tilde{i}_\beta + \frac{L_d - L_q}{L_d}\hat{\omega}_e\tilde{i}_\alpha - \frac{1}{L_d}\tilde{e}_\beta - \frac{1}{L_d}k\mathrm{sgn}\left(\tilde{i}_\beta\right)\right\}\\ =&-\frac{R_s}{L_d}\left(\tilde{i}_\alpha^2 + \tilde{i}_\beta^2\right) - \frac{1}{L_d}\left\{\tilde{i}_\alpha\left(\tilde{e}_\alpha + k\mathrm{sgn}\left(\tilde{i}_\alpha\right)\right)\right\} - \frac{1}{L_d}\left\{\tilde{i}_\beta\left(\tilde{e}_\beta + k\mathrm{sgn}\left(\tilde{i}_\beta\right)\right)\right\}\\ =&V_1 - V_2\end{aligned} \tag{1.48}$$

where

$$V_1 = -\frac{R_s}{L_d}\left(\tilde{i}_\alpha^2 + \tilde{i}_\beta^2\right), \tag{1.49}$$

$$\begin{aligned}V_2 =& \frac{1}{L_d}\left\{\tilde{i}_\alpha\left(\tilde{e}_\alpha + k\mathrm{sgn}\left(\tilde{i}_\alpha\right)\right)\right\} + \frac{1}{L_d}\left\{\tilde{i}_\beta\left(\tilde{e}_\beta + k\mathrm{sgn}\left(\tilde{i}_\beta\right)\right)\right\}\\ =&\frac{1}{L_d}\left\{\tilde{i}_\alpha\left(\frac{s + 2\omega_{cl}}{s + \omega_{cl}}k\mathrm{sgn}\left(\tilde{i}_\alpha\right) - e_\alpha\right) + \tilde{i}_\beta\left(\frac{s + 2\omega_{cl}}{s + \omega_{cl}}k\mathrm{sgn}\left(\tilde{i}_\beta\right) - e_\beta\right)\right\}\end{aligned}. \tag{1.50}$$

Equation (1.49) shows that V_1 is always negative. The LPF cut-off frequency is relatively high, $\frac{s+2\omega_{cl}}{s+\omega_{cl}} \approx 2$. If k is large enough and satisfy (1.51),

$$k > \frac{1}{2}\max\left(|e_\alpha|,\ |e_\beta|\right), \tag{1.51}$$

the sliding mode convergence condition can be satisfied. When V converges to zero, $\tilde{i}_\alpha$ and $\tilde{i}_\beta$ will be zero and the estimated currents will converge to the actual values.

By using the idea of equivalent control, the stability of the full-order sliding mode observer can also be analyzed.

Step 1: Refering to the stability analysis of the traditional reduced-order sliding mode observer, as shown in (1.45)–(1.51). The difference is that the full-order sliding mode observer does not need the low pass filter which causes the phase lag in the rotor position observation. However, its derivation process is basically the same.

Step 2: When the stator currents reach the sliding mode region, the estimated stator currents converge to their actual values, i.e. $\dot{\tilde{i}}_\alpha = \dot{\tilde{i}}_\beta = 0$. At this time, (1.46) can be rewritten as follows,

$$\begin{aligned}\tilde{e}_\alpha &= -k\mathrm{sgn}\left(\hat{i}_\alpha - i_\alpha\right)\\ \tilde{e}_\beta &= -k\mathrm{sgn}\left(\hat{i}_\beta - i_\beta\right)\end{aligned}. \quad (1.52)$$

Let (1.44) be subtracted from (1.43), the error dynamic equation of EEMF can be obtained as:

$$\begin{aligned}\dot{\tilde{e}}_\alpha &= -\tilde{e}_\beta\hat{\omega}_e + \frac{1}{L_d}m\mathrm{sgn}\left(\hat{i}_\alpha - i_\alpha\right)\\ \dot{\tilde{e}}_\beta &= \tilde{e}_\alpha\hat{\omega}_e + \frac{1}{L_d}m\mathrm{sgn}\left(\hat{i}_\beta - i_\beta\right)\end{aligned}. \quad (1.53)$$

According to (1.52) and (1.53), the results can be obtained,

$$\begin{aligned}\dot{\tilde{e}}_\alpha &= -\tilde{e}_\beta\hat{\omega}_e - \frac{1}{L_d}m/k\cdot\tilde{e}_\alpha\\ \dot{\tilde{e}}_\beta &= \tilde{e}_\alpha\hat{\omega}_e - \frac{1}{L_d}m/k\cdot\tilde{e}_\beta\end{aligned}. \quad (1.54)$$

Equation (1.54) shows that the EEMF error dynamic equations have two basic terms: the prediction and the correction term. EEMF observation equations are the prediction term, and the feedback correction composed of the gain coefficient $\frac{1}{L_d}m/k$ is the correction term. Therefore, the above-mentioned full-order sliding mode observer can remove the LPF used in the traditional reduced-order sliding mode observer, thereby avoiding the phase lag.

Solve the differential (1.54) and obtain its characteristic equation,

$$\alpha(s) = s^2 + 2/L_d\cdot m/k\cdot s + \left(\hat{\omega}_e^2 + 1/L_d^2\cdot(m/k)^2\right) = 0 \quad (1.55)$$

where 's' is Laplace operator. The eigenvalue can be given as,

$$s_{1,2} = \frac{-m/k \pm jL_d\omega_e}{L_d}. \quad (1.56)$$

Equation (1.56) shows that the eigenvalues s_1 and s_2 are conjugate complex roots, which are located on the left half plane. And the system is asymptotically stable.

Equation (1.55) can be further rewritten as the following standard form,

$$s^2 + 2\xi\omega_n s + \omega_n^2 = 0 \quad (1.57)$$

where

$$\omega_n = \sqrt{\hat{\omega}_e^2 + 1/L_d^2\cdot(m/k)^2}$$

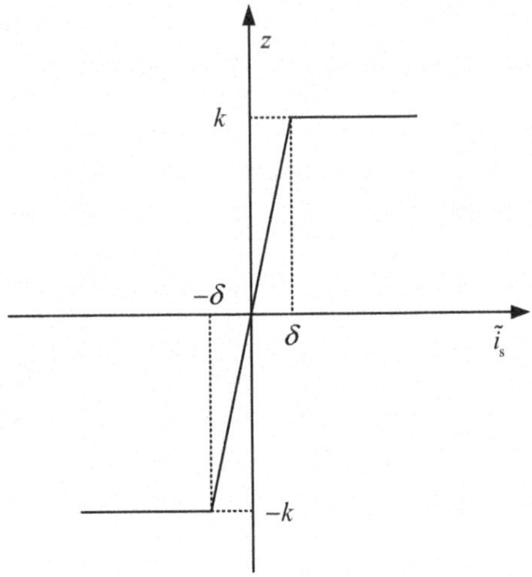

Fig. 1.18 Characteristic curve of saturation function

$$\xi = \frac{m/k}{\sqrt{L_d^2\hat{\omega}_e^2 + (m/k)^2}}. \tag{1.58}$$

where ω_n refers to the natural oscillation angular frequency, and ξ is the damping ratio.

According to the control theory, the convergence rate of the EEMF observation errors $\tilde{e}_{s\alpha}$ and $\tilde{e}_{s\beta}$ depend on $\xi\omega_n$, i.e. the value of $1/L_d{\cdot}m/k$. The larger the value of $\xi\omega_n$ is, the faster the convergence of the EEMF observation error will be. For $\xi \in (0.4, 0.8)$, the overshoot of the system is suitable, with a shorter adjustment time and better dynamic performance. In order to diminish the chattering phenomenon inherently in the sliding mode observer, the saturation function is used to replace the traditional sign function, as shown in Fig. 1.18

$$z_{\alpha,\beta} = \begin{cases} k, & \tilde{i}_s \geq \delta \\ k \cdot \tilde{i}_s / \delta, & -\delta < \tilde{i}_s < \delta \\ -k, & \tilde{i}_s \leq -\delta \end{cases} \tag{1.59}$$

where δ is the boundary layer constant of current error. The saturation function is a quasi-sliding mode observer method, which is essentially a linear control in the boundary layer and a non-continuous switching control outside the boundary layer. Therefore, this method could effectively weaken the chattering phenomenon.

1.6 Summary

The basic knowledge of PMSM drives has been introduced in this section. The voltage equations of the PMSM in the *a-b-c* coordinate frame are severely coupled, which is mainly caused by the mutual inductances. In order to realize the decoupling control, the α-β and the *d-q* coordinate frames are introduced, which aim to eliminate the effect of the mutual inductances. The PMSM models in the α-β and *d-q* coordinate frames are obtained by the coordinate transformation, which are an important tool for PMSM drives. The control methods of PMSM based on vector control are then introduced, including the SVPWM strategy, the corresponding maximum voltage vector evaluation and field orientation control. Nowadays, as the expanded usage of PMSMs in industry and home appliance, higher requirements of motor drives are proposed, such as lower cost and higher reliability. Therefore, position sensorless control method becomes an attractive researching area. This chapter introduces the model-based methods to achieve sensorless control scheme for IPMSM drives. The basic principle of sliding-mode observer based sensorless control is introduced.

References

1. T. Wildi, *Electrical Machines, Drives, and Power Systems*, 2nd edn. (Prentice-Hall, Englewood Cliffs, NJ, 1991)
2. P.C. Krause, *Analysis of Electric Machinery* (McGraw-Hill, New York, 1987)
3. A.E. Fitzgerald, C. Kingslay, S.D. Umans, *Electric Machinery* (McGraw-Hill, New York, 1983)
4. S. Bolognani, M. Zigliotto, Novel digital continuous control of SVM inverters in the overmodulation range. IEEE Trans. Ind. Appl. **33**(2), 525–530 (1997)
5. F. Briz, A. Diez, M. W. Degner, R. D. Lorenz, Current and flux regulation in field-weakening operation [of induction motors]. IEEE Trans. Ind. Appl. **37**(1), 42–50 (2001)
6. Z. Chen, M. Tomita, S. Doki, S. Okuma, An extended electromotive force model for sensorless control of interior permanent-magnet synchronous motors. IEEE Trans. Indu. Electron. **50**(2), 288–295 (2003)

Chapter 2
High Power Factor Control of Grid Input Current

Large volume electrolytic capacitors are usually utilized at the DC-link of PMSM drives to buffer and store energy, which can stabilize the DC-link voltage and supply power for the inverter. Meanwhile, the power factor correction (PFC) circuit is necessary to improve the power factor and reduce the total harmonic distortion (THD) of the grid input current. However, the electrolytic capacitor containing the electrolyte results in lifetime decrease and volume increase of the drive system [1–4]. In order to solve these problems, the reduced DC-link capacitance AC motor drive technique has been developed [5–8]. Compared with the convensional drive system, the decreased capacity of the DC-link capacitor can effectively reduce the drive system volume. Meanwhile, the cost of the system can also be reduced, which is an important concern in industry applications.

Due to the remarkably decreased capacity of the DC-link capacitor, the energy stored in the film capacitor is difficult to maintain DC-link voltage constant. As a result, the DC-link voltage will fluctuate with the grid voltage. However, as the constant DC-link voltage in the traditional motor drive system leads to a narrow conduction angle of the diode rectifier, the PFC circuit is required to broaden the conduction angle and reduce the current harmonics. In this way, the fluctuated DC-link voltage makes it possible to achieve higher input power factor. Therefore, the PFC circuit can be eliminated to reduce the cost and improve the reliability of the drive system. Hence the high power factor control strategy of the reduced DC-link capacitance motor drive system is important.

© The Author(s), under exclusive license to Springer Nature Singapore Pte Ltd. 2021
G. Wang et al., *Reduced DC-link Capacitance AC Motor Drives*,
https://doi.org/10.1007/978-981-15-8566-1_2

2.1 Power Characteristic Analysis of Drive System

2.1.1 *Topology of Single-Phase Reduced DC-Link Capacitance Motor Drives*

As shown in Fig. 2.1, the reduced DC-link capacitance single-phase PMSM drive system consists of the diode rectifier, the DC-link film capacitor and the inverter. The capacity of the applied DC-link electrolytic capacitor is selected from several hundred to several thousand microfarads in the drive system using a single-phase diode rectifier. In the electrolytic capacitor-less drive system, the capacity is usually selected as 2–5% of the electrolytic capacitor. As a result, the DC-link voltage will fluctuate with the grid voltage. The grid input power and the inverter output power also fluctuate with the DC-link voltage.

2.1.2 *Grid Input Power*

Figure 2.2 shows the simplified block diagram of the electrolytic capacitor-less motor drive system, the grid voltage u_g can be presented as

$$u_g = U_g \sin(\omega_g t + \varphi_g) \tag{2.1}$$

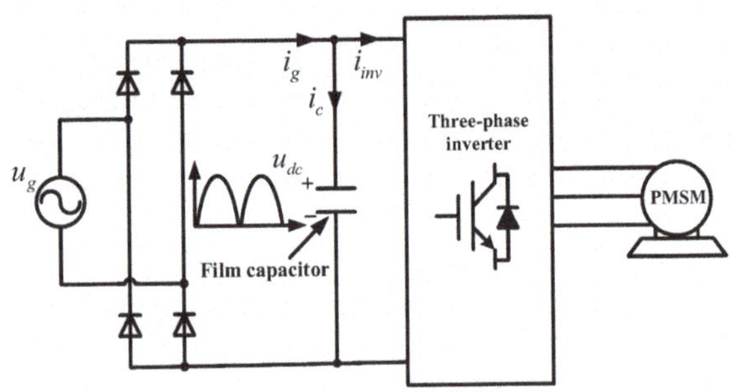

Fig. 2.1 Reduced DC-link capacitance single-phase PMSM drive

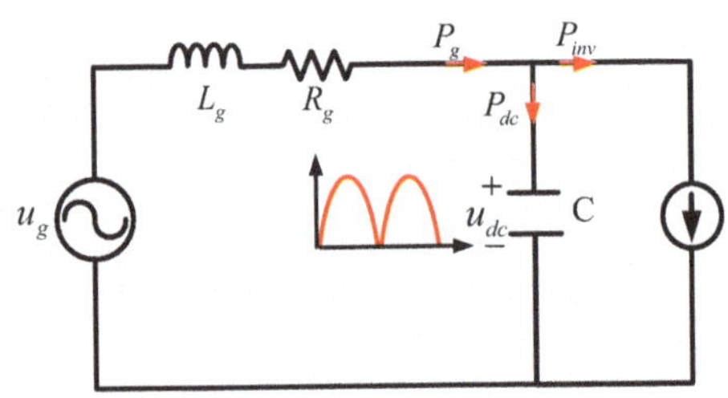

Fig. 2.2 Simplified block diagram of reduced DC-link capacitance PMSM drive system

where U_g, ω_g and φ_g denote the peak amplitude, the angular velocity, and the initial phase of the grid voltage, respectively. In order to achieve the unity power factor, the phase of the grid current i_g should be synchronous with the grid voltage. Hence the expected grid input power P_g can be calculated as

$$P_g = u_g i_g = U_g I_g \sin^2(\omega_g t + \varphi_g) \tag{2.2}$$

where I_g is the amplitude of the grid current.

2.1.3 Inverter Output Power

The grid input power of the drive system P_g can be divided into two parts: the film capacitor power P_{dc} and the three-phase inverter power P_{inv}, which can be expressed as

$$P_g = P_{dc} + P_{inv}. \tag{2.3}$$

Compared with the traditional drive system, the small volume of the film capacitor cannot maintain the DC-link voltage constant. The DC-link voltage will fluctuate with twice the frequency of grid voltage and it can be expressed as

$$\begin{aligned} u_{dc} &= \left|U_g \sin(\omega_g t + \varphi_g)\right| \\ &= U_g \sin(\omega_g t + \varphi_g) \cdot sgn(\sin(\omega_g t + \varphi_g)). \end{aligned} \tag{2.4}$$

As a result, P_{dc} can be calculated as follows:

$$\begin{aligned} P_{dc} &= u_{dc} i_c - u_{dc} C_{dc} \frac{du_{dc}}{dt} \\ &= u_{dc} C_{dc} \frac{d}{dt} \{U_g \sin(\omega_g t + \varphi_g) \cdot sgn(\sin(\omega_g t + \varphi_g))\} \\ &= \frac{1}{2} U_g^2 C_{dc} \omega_g \sin(2(\omega_g t + \varphi_g)) \end{aligned} \tag{2.5}$$

where i_c and C_{dc} are the current and capacitance of the film capacitor, sgn() is the sign function. Ignoring the switching loss of the inverter, the motor power can be regarded as the inverter power P_{inv}. Therefore, the inverter power P_{inv} can be calculated as:

$$P_{inv} = 1.5\left(\vec{u}_{dq} \cdot \vec{i}_{dq}\right) = 1.5(u_d i_d + u_q i_q) \tag{2.6}$$

$$u_d = R_s i_d + L_d \frac{di_d}{dt} - \omega_e L_q i_q \tag{2.7}$$

$$u_q = R_s i_q + L_q \frac{di_q}{dt} + \omega_e L_d i_d + \omega_e \psi_f, \tag{2.8}$$

2.2 Inverter Power Control

2.2.1 Principle of Inverter Power Control

As shown in Fig. 2.3, the grid input power P_g, the DC-link capacitor power P_{dc} and the inverter output power P_{inv} fluctuate with twice the frequency of grid voltage. Based on the power analysis of the drive system, the solution to achieve high power factor and low THD of the grid current is to ensure the grid input power to be sinusoidal. As shown in Fig. 2.3, the grid input power depends on both the power of the film capacitor and the inverter. The film capacitor power can be calculated by the fluctuated DC-Link voltage according to (2.5), which cannot be controlled directly. Meanwhile, the sum of the film capacitor power per cycle is zero, and the magnitude of the power is much smaller than the inverter power. As a result, the approach to control the grid input power is to regulate the inverter power effectively.

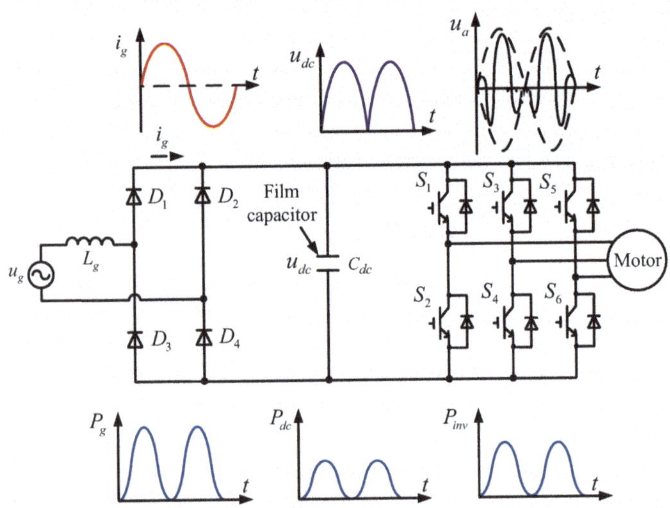

Fig. 2.3 Power characteristics of the drive system

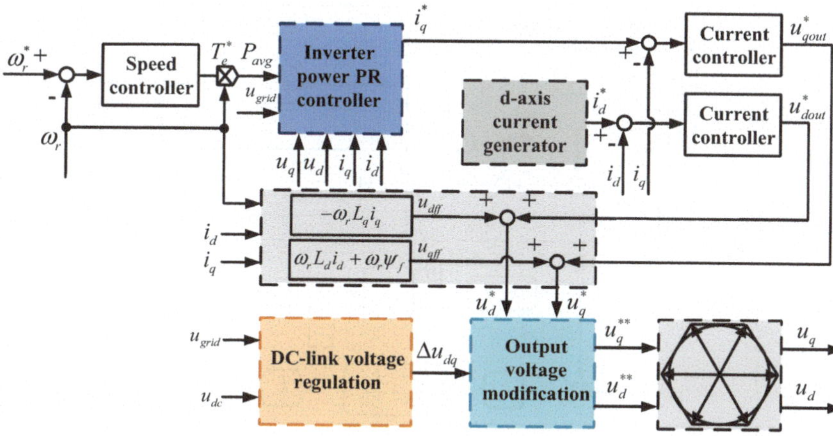

Fig. 2.4 Block diagram of inverter power control

2.2.2 Inverter Power Control Scheme

Figure 2.4 shows the block diagram of the proposed inverter power control strategy. The proportional resonant (PR) controller is used to regulate the inverter power, which mainly generates the q-axis current reference. However, the inverter power loop cannot control the inverter power perfectly. Here, a power compensation method based on the DC-link voltage regulation is adopted to diminish the power control error. Meanwhile, the output voltage modification block aims to impose the output of the DC-link voltage regulation to the decoupled voltage vectors. The generation of the current reference for flux weakening should take the DC-link voltage fluctuation into consideration. Additionally, as the over-modulation of SVPWM distorts the output voltage of inverter. Then the stator current will contain remarkable ripples, which will lead to the distortions of the grid current. In order to reduce the harmonics of the grid current, the proposed SVPWM method ensures that the magnitude of the voltage vector satisfies the maximum output voltage circle of the inverter.

2.2.3 Inverter Power Control Loop

Figure 2.5 shows the block diagram of the inverter power controller. As shown in (2.3), the grid input power consists of the film capacitor power and the inverter power. According to (2.5), the average power of the film capacitor is zero, and the average grid input power is equal to the average inverter power. The average inverter power reference P_{avg} is regulated by the product between the torque reference T_e^* and the angular speed ω_r. The grid voltage phase θ_g is detected by a phase locked loop (PLL).

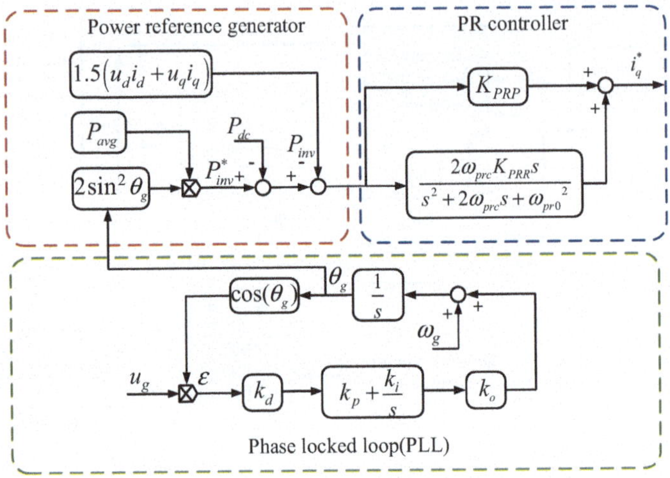

Fig. 2.5 Block diagram of the inverter power controller

Then P_{avg} is regulated by a sinusoidal waveform to generate the grid input power reference P^*_{inv}, which fluctuates with twice the frequency of the grid voltage.

Actually, the DC-link voltage cannot decrease to zero when the load is not large enough, and the typical waveform of the DC-link voltage is shown in Fig. 2.6. The film capacitor power P_{dc} is related to the DC-link voltage and can be calculated as follows:

$$P_{dc} = \begin{cases} 0 & \theta \in [0,\, \theta_m) \cup (\pi - \theta_m,\, \pi] \\ \frac{1}{2}U_g^2 C_{dc}\omega_g \sin(2\theta_g) & \theta \in [\theta_m,\, \pi - \theta_m] \end{cases} \tag{2.9}$$

where θ_g and θ_m denote the phase of the grid voltage and the initial conduction phase of the grid current, respectively.

The inverter power control loop is established by regarding the calculation of the motor output power as the power feedback P_{inv}, which also represents the mechanical load of the motor. It can be seen that the power signals in the inverter power control loop are the periodic sinusoidal signals. Hence both the repetitive controller and PR controller can achieve better performance than the PI controller which is effective

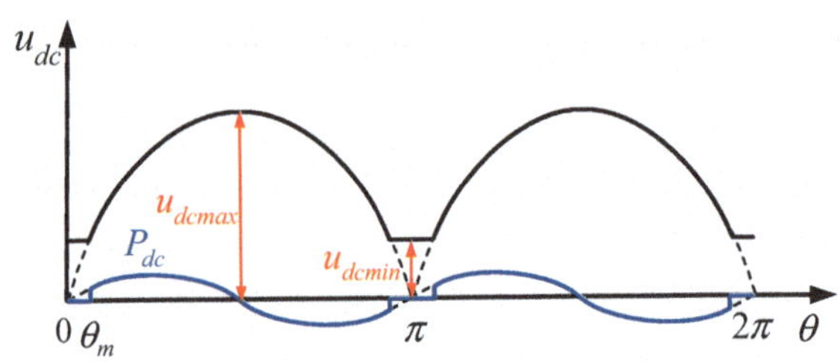

Fig. 2.6 Typical waveform of the DC-link voltage

to the DC signals. The repetitive controller and the PR controller are based on the internal model theory, which can realize high precision control for the feedback system by specific controllers combined with the internal model of input signal or disturbance signal. The repetitive controllers are applied successfully to track the periodic input signal or reject the periodic disturbance. As for the sinusoidal signals, the PR controller containing the corresponding internal model is applied to optimize the control performance.

Both the repetitive controller and the PR controller are fit for the periodic signals. The repetitive controller needs several periods to learn the signal model and eliminate the errors by the sum of errors in the previous periods. As for the PR controller, the signal model has been previously set based on the internal model of the sinusoidal signal. Therefore, as for the sinusoidal signals, the control performance of PR controller can be better than that of the repetitive controller. Meanwhile, the repetitive controller should be combined with a PI controller to satisfy the basic system requirements on closed-up bandwidth, loop gain, etc. More parameters need to be designed than that of PR controller. As a result, the parameter determination and system stability consideration applying repetitive controller will be trivial compared to the PR controller. Hence the PR controller is applied in the proposed inverter power control loop to achieve better performance and a simple control block, which is easy to determine the parameters.

2.3 Parameter Determination of Inverter Power Controller

2.3.1 *Mathematical Model of Inverter Power Control Loop*

The PR controller can be expressed as follows:

$$\mathrm{G}_{PR}(\mathrm{s}) = K_{PRP} + K_{PRR}\frac{2\omega_{prc}s}{s^2 + 2\omega_{prc}s + \omega_{pr0}^2} \tag{2.10}$$

where K_{PRP} and K_{PRR} are the proportional and the resonant gains, respectively, ω_{pr0} is the resonant frequency, and ω_{prc} is the cut-off frequency. The output of PR controller depends on the error between the reference and the actual values of inverter power, which can be regarded as the q-axis current reference. For the steady state, the average q-axis current is proportional to the load torque and the instantaneous value will fluctuate with twice the frequency of the grid voltage. Therefore, the output of PR controller contains the twice frequency component and the high frequency component for the stator current regulation. The values of K_{PRP} and K_{PRR} can be determined as follows [9–11].

The power control loop using the PR controller can be shown in Fig. 2.7, where T_{prd} is the time delay of the control loop and the motor can be simplified as a *RL* circuit. The open loop transfer function can be expressed as

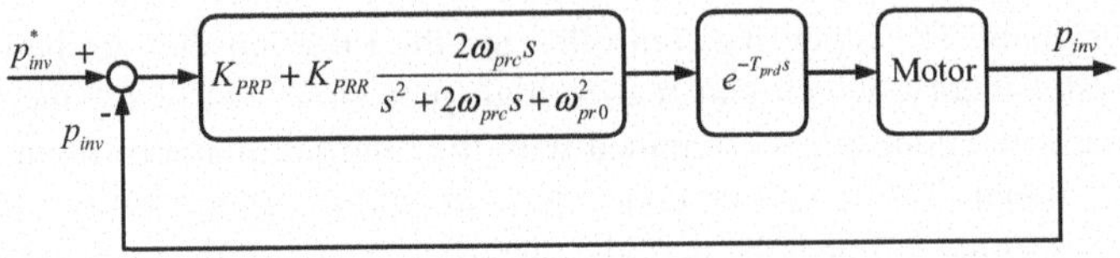

Fig. 2.7 Power control loop of PR controller

$$G_{PRL}(s) = K_{PRP}\left(1 + \frac{2\omega_{prc}s}{T_{PRP}\left(s^2 + 2\omega_{prc}s + \omega_{pr0}^2\right)}\right) \cdot e^{-T_{prd}s} \cdot \frac{1}{R_{prr}\left(1 + sT_{pri}\right)} \tag{2.11}$$

where $T_{PRP} = K_{PRP}/K_{PRR}$ and $T_{pri} = L_{prl}/R_{prr}$.

2.3.2 Parameters Design of PR Controller

The phase angle of the open loop transfer function $G_{PRL}(s)$ at the cross over frequency ω_{prf} is given by

$$\begin{aligned}\angle G_{PRL}\left(j\omega_{prf}\right) =& \angle\left\{\frac{K_{PRP}}{R_{prr}}\left(1 + \frac{2j\omega_{prf}\omega_{prc}}{T_{PRP}\left\{\left(\omega_{pr0}^2 - \omega_{prf}^2\right) + 2j\omega_{prf}\omega_{prc}\right\}}\right)\frac{e^{-j\omega_{prf}T_{prd}}}{1 + j\omega_{prf}T_{pri}}\right\} \\ =& \left(-\pi + \varphi_{prm}\right) \\ \approx& \tan^{-1}\left(\omega_{prf}T_{PRP}/2\omega_{prc}\right) - \pi/2 - \omega_{prf}T_{prd} - \tan^{-1}\left(\omega_{prf}T_{pri}\right)\end{aligned} \tag{2.12}$$

where φ_{prm} represents the required stability phase margin which is usually taken as 40° in theory. The phase margin can be represented as

$$\varphi_{prm} \approx \tan^{-1}\left(\omega_{prf}T_{PRP}/2\omega_{prc}\right) - \omega_{prf}T_{prd}. \tag{2.13}$$

The maximum value of ω_{prf} for a given φ_{prm} can be represented as

$$\omega_{prf\,(max)} = \frac{\pi/2 - \varphi_{prm}}{T_{prd}}. \tag{2.14}$$

The maximum magnitude of K_{PRP} can be achieved by setting the open loop gain at this frequency $\omega_{prf(\max)}$ to unity, which gives

$$K_{PRP} = R_{prr}\omega_{prf\,(\max)}T_{PRP}\frac{\sqrt{1 + \left(\omega_{prf\,(\max)}T_{pri}\right)^2}}{\sqrt{\left(\omega_{prf\,(\max)}T_{PRP}\right)^2 + \left(2\omega_{prc}\right)^2}}. \tag{2.15}$$

Therefore, the magnitude of K_{PRP} can be represented as

$$K_P = \omega_{f(\max)} L. \tag{2.16}$$

The magnitude of K_{PRR} can be maximized by making

$$\tan^{-1}\left(\omega_{prf(\max)} T_{PRP}/2\omega_{prc}\right) \approx \pi/2 \tag{2.17}$$

which gives

$$T_{PRP} = \frac{20\omega_{prc}}{\omega_{prf(\max)}}. \tag{2.18}$$

The values of K_{PRP} and K_{PRR} are determined by the above method when the equivalent *RL* circuit of the power control loop is regarded as the *q*-axis *RL* circuit. In the drive system, the coupled inverter power between *d*- and *q*-axis leads the *d*-axis inductance to affect the accuracy of the equivalent *RL* circuit. The parameters of PR controller need to be tuned in practical applications to optimize the control performance. Fortunately, the tuning is relatively simple since the values of the *d*- and *q*-axis inductances are in the same order of magnitude.

2.3.3 *Parameters Determination*

As the frequency of the grid voltage is 50 Hz and the DC-link voltage rectified by the single-phase diode rectifier fluctuates with 100 Hz, the resonant frequency ω_{pr0} should be set as 100 Hz. Meanwhile, the PR controller can achieve better performance when the cut-off frequency ω_{prc} is set lower to obtain a higher gain for the inverter power reference. However, it should be noted that the lower value of ω_c may result in the instability of the drive system for the sharper phase distortion. The Bode diagram of the PR controller is shown in Fig. 2.8.

2.4 Inverter Power Compensation Based on DC-Link Voltage Control

2.4.1 *Performance Evaluation of Inverter Power Control*

The inverter power controller is introduced in Sect. 2.2. While the error between the power reference P_{inv}^* and the actual feedback power P_{inv} is difficult to be eliminated only by adopting the PR controller in the inverter power control loop. The PR controller can obtain better performance for the fluctuated inverter power component.

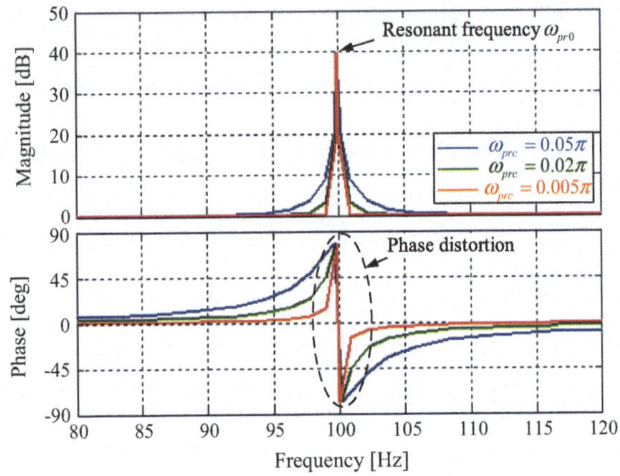

Fig. 2.8 Bode diagram of PR controller of different ω_c values

However, as for the fluctuated current signals, the traditional PI current controllers limit the control performance. Meanwhile, the inverter power is coupled between *d*- and *q*-axes.

As the feedback of the inverter power only has an effect on i_q reference and the regulation of i_d does not depend on the inverter power control loop, the reference of i_q generated from the PR controller is not the optimal matching. Once taking the relationship of the coupled inverter power in the *d*-*q* axis reference frame into consideration, the accurate regulation of i_d is not easy to realize. As a result, it is also difficult to separate the coupled power between the *d*- and *q*-axis completely.

Therefore, it is necessary to adopt an additional inverter power compensation block to reduce the inverter power control error, which can realize the actual power feedback tracking the power reference better. Meanwhile, the bandwidth of current PI controllers can be set lower, which benefits the stability of the drive system.

2.4.2 Closed Loop Control of DC-Link Voltage Control

A novel power compensation method based on the DC-link voltage regulation is illustrated in this section. The traditional drive system needs a PFC circuit to ensure that the grid current can satisfy the standard requirements of EN-61000-3-2. The PFC circuit also blocks the grid input side and the inverter output side. Due to the elimination of the PFC circuit in the reduced DC-link capacitance motor drive system, the grid input side and inverter output side have been connected by the DC-link directly. The distorted grid input current caused by the distorted inverter output power is reflected by the DC-link voltage. Hence the distorted inverter power can be detected by the DC-link voltage and the additional power compensation can be realized by the DC-link voltage regulation.

According to the circuit topology of Fig. 2.1, the inverter power can be calculated as:

$$P_{inv} = u_{dc}i_{inv} = u_{dc}\left(i_g - C_{dc}\frac{du_{dc}}{dt}\right) \tag{2.19}$$

where i_{inv} represents the input current of the inverter.

Generally, the DC-link voltage can be calculated as shown in (2.4). However, if the inverter power cannot be controlled precisely, the irregular inverter power will distort the DC-link voltage. The Fourier analysis of the actual DC-link voltage can be expressed as:

$$u_{dc} = u_{dc,0} + \sum_{k=1}^{n} u_{dc,k}\sin(2k\omega_g t + \varphi_{dc}) \tag{2.20}$$

where $u_{dc,0}$, $u_{dc,k}$ and φ_{dc} are the offset value, the amplitudes, and the phases of the kth harmonic components of DC-link voltage, respectively.

Therefore, the inverter power control error can be detected by the distortion of the DC-link voltage as shown in Fig. 2.9a. The power compensation method based on the DC-link voltage regulation is applied to improve the inverter power control performance, which is shown in Fig. 2.9b. The DC-link voltage reference is generated by the sine calculation of the estimated phase of the grid voltage. A PI controller is applied to regulate the DC-link voltage. The output of the PI controller represents the power error ΔP that can be calculated as follows:

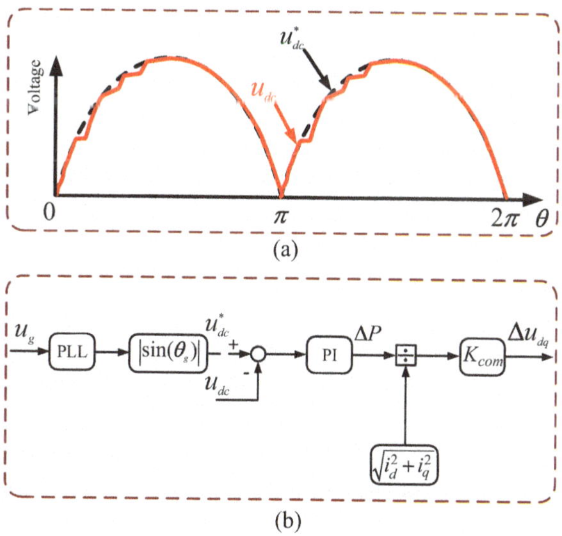

Fig. 2.9 Scheme of the inverter power compensation. **a** The DC-link voltage reference and the actual distorted DC-link voltage. **b** The block diagram of the DC-link voltage regulation

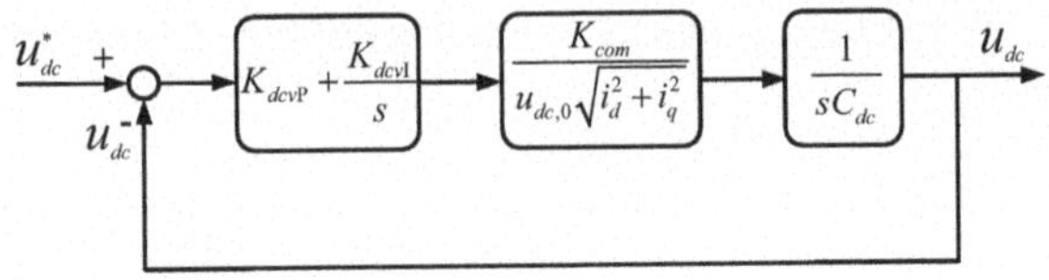

Fig. 2.10 Small signal block diagram of the DC-link voltage control

$$\Delta P = \left(K_{dcvP} + \frac{K_{dcvI}}{s}\right)\left(u_{dc}^* - u_{dc}\right) \tag{2.21}$$

where K_{dcvP} and K_{dcvI} represent the proportional gain and the integral gain of the DC-link controller, u_{dc}^* and u_{dc} represent the DC-link voltage reference and the actual DC-link voltage, respectively.

Then, ΔP is divided by the magnitude of the current vector, which represents the magnitude of the modified voltage Δu_{dq} in the d-q axis reference frame. Actually, the modified voltage Δu_{dq} will lead to the additional motor power ΔP, which can affect Δi_{inv}. The relationship can be given by

$$\Delta i_{inv} = \frac{\Delta P}{u_{dc,0}}. \tag{2.22}$$

A reactor is applied to improve the quality of the grid current in the reduced DC-link capacitance motor drive system, which can stabilize the grid current and reduce the impact caused by Δi_{inv}. As a result, the current of the film capacitor Δi_c is regulated by Δi_{inv}, which can shape the DC-link voltage Δu_{dc},

$$\Delta u_{dc} = \frac{1}{C_{dc}} \int \Delta i_c(t)dt. \tag{2.23}$$

Therefore, the small signal block diagram of the DC-link voltage regulation can be simply expressed as Fig. 2.10. The establishment of the DC-link control loop can explain the relationship between the motor power and the DC-link voltage, which can shape the DC-link voltage effectively. Meanwhile, the control loop can be regarded as a first-order system, and it is easy to design the parameters of PI controller. Therefore, the DC-link voltage regulation can effectively eliminate the distorted inverter power and improve the power control performance.

2.4.3 *DC-Link Voltage Reference Generation*

It can be seen from Fig. 2.11 that the minimum DC-link voltage u_{dc_min} is closely related to the conduction angle of the diode rectifier, which has an important effect on the grid current harmonics. The maximum DC-link voltage u_{dc_max} is determined by the grid voltage. The grid current shown in Fig. 2.11 can be approximately represented as:

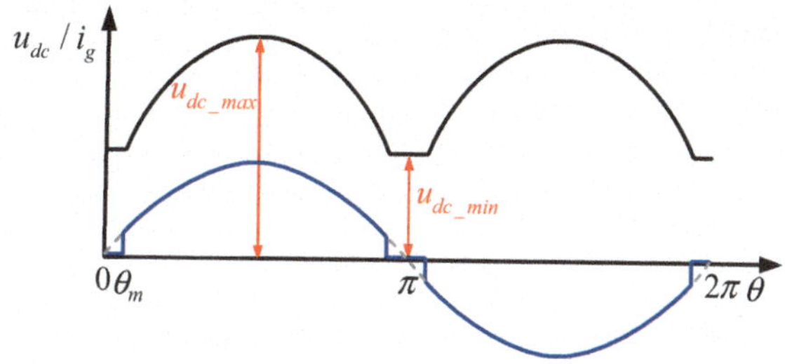

Fig. 2.11 Relationship between the grid current and the fluctuated DC-link voltage

$$i_g = \begin{cases} I_g \sin\theta, & \theta \in [\theta_m, \pi - \theta_m] \cup [\pi + \theta_m, 2\pi - \theta_m] \\ 0, & \theta \in [0, \theta_m] \cup [\pi - \theta_m, \pi + \theta_m] \cup [2\pi - \theta_m, 2\pi] \end{cases}. \tag{2.24}$$

As the function of i_g can be regarded as an odd one, the Fourier analysis of i_g is the sinusoidal series, which can be calculated as:

$$\begin{aligned} i_g &= \sum_{n=1}^{\infty} b_n \sin(nx) \\ b_n &= \frac{2}{\pi} \int_{\theta_m}^{\pi-\theta_m} I_g \sin\theta \sin(n\theta) d\theta, n = 1, 3, 5, 7, \cdots \end{aligned} \tag{2.25}$$

The Fourier coefficients b_n can be presented as:

$$b_n = \frac{I_g}{\pi}\left(\frac{1}{n+1}\sin((n+1)\theta_m) - \frac{1}{n-1}\sin((n-1)\theta_m)\right). \tag{2.26}$$

It can be seen that b_n is related to I_g. For example, if the value of I_g is unit, the grid current harmonics with respect to θ_m can be shown in Fig. 2.12.

The other orders of the grid current harmonics can be calculated by (2.26). As a result, the value of θ_m can be obtained to satisfy the harmonic standards, which are shown in Table 2.1. Hence u_{dc_min} can be determined according to the value of grid currents, which can make the system satisfy the harmonic standards. The detailed values under different amplitude of grid currents can be shown in Fig. 2.13. It can be seen that the DC-link voltage reference u_{dc_min} decreases as the grid current increases.

2.4.4 DC-Link Voltage Control Realization

As shown in Fig. 2.4, the outputs of the current controllers are decoupled by the feed-forward voltage u_{dff} and u_{qff} to generate the voltage vector reference u_{dq}^*. Whereas the output of the DC-link voltage regulation shown in Fig. 2.9 needs to be imposed to the final voltage vector reference. The purpose of the output signal Δu_{dq} is to

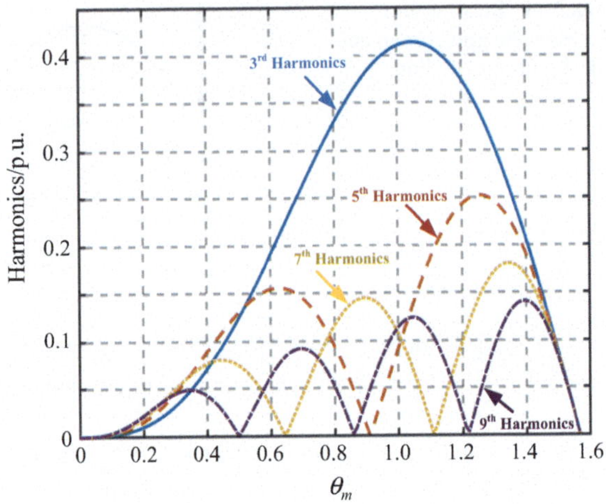

Fig. 2.12 Harmonics of grid current with respect to the initial conduction phase θ_m

Table 2.1 Harmonic Standards of EN61000-3-2

Harmonics[nth]	Standards[A]	Harmonics[nth]	Standards[A]
2	1.08	3	2.30
4	0.43	5	1.14
6	0.3	7	0.77
8	0.23	9	0.40
10	0.18	11	0.33
12	0.15	13	0.21
Even harmonics ($14 \leq n \leq 40$)	1.84/n	Odd harmonics ($15 \leq n \leq 39$)	2.25/n

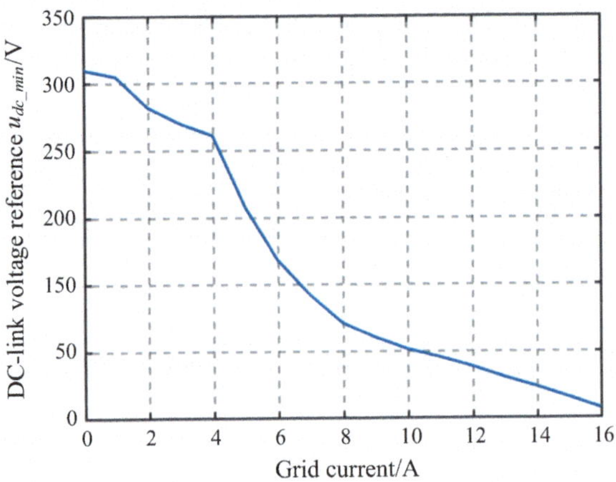

Fig. 2.13 Determination of DC-link voltage reference with respect to the altitude of grid current

modify the inverter power, which also can be regarded as a voltage disturbance for the closed-loop control system.

Therefore, the smaller the magnitude of the imposed voltage vector is, the less the effect of the voltage disturbance will be on the drive system. The additional effect of the voltage vector on the motor current has been considered in the voltage modification module in [7]. Hence the optimized voltage modification vectors are the ones with the shortest distance between the final voltage vector and the feedforward voltage vector. In the proposed output voltage modification method, the feed-forward voltage vector and the output of current controller are treated as a whole voltage vector u^*_{dq}, which owns the smallest magnitude for certain power modification, because the effect of the additional voltage vector on the motor current is not obvious. Hence the compromise between the consideration of the effect on the motor current in [7] and the reduced calculation of the proposed method is tolerable.

The imposed voltage vector in the *d-q* reference frame is shown in Fig. 2.14a, and the possible direction of voltage vector is infinite, which means the value of K_{com} as shown in Fig. 2.10 is various. However, the direction paralleled to the current vector is the optimization, because the magnitude of the voltage is the smallest one to the

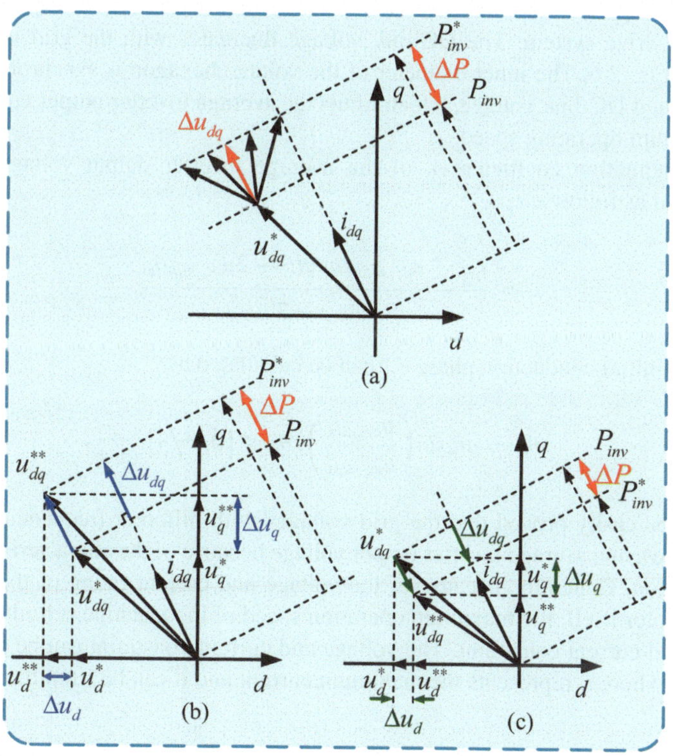

Fig. 2.14 Scheme of the output voltage modification. **a** Selection of the modified voltage direction. **b** $P_{inv} < P^*_{inv}$. **c** $P_{inv} > P^*_{inv}$

same inverter power modification. Meanwhile, K_{com} can be regarded as a unity gain. Figure 2.14b shows the scheme of the synthetic voltage vector modification when the inverter power error $\Delta P > 0$. It means that P_{inv} is smaller than P^*_{inv}. In the figure, Δu_d and Δu_q are the d- and q-axis modification voltages, respectively. By applying the voltage modification block, the final voltage reference u^{**}_{dq} can decrease the inverter power error effectively. Figure 2.14c shows the situation that $\Delta P < 0$ and the direction of Δu_{dq} is opposite to the one in Fig. 2.14b in order to decrease the inverter power. As a result, the performance of inverter power control can be improved by applying the modified voltage.

2.4.5 Analysis of Maximum Motor Speed

The motor speed limitation is related to the grid voltage, the DC-link voltage, and the load torque. The drive system should meet the voltage and current constraint to guarantee the normal motor operation. As for the reduced DC-link capacitance motor drive system, the maximum operating speed will be determined by the fluctuated DC-link voltage for the remarkably reduced DC-link capacitance compared to the traditional drive system. The DC-link voltage fluctuates with the grid voltage as shown in Fig. 2.6. The inner diameter of the voltage hexagon is synchronous with the fluctuated DC-link voltage, which limits the average inverter output voltage and the maximum operating speed.

The attenuation coefficient λ of the average inverter output voltage can be represented as follows:

$$\lambda = \frac{\int_{\theta_m}^{\pi-\theta_m} u_{dc\_max} sin\theta d\theta + 2u_{dc\_min}\theta_m}{u_{dc\_max}\pi} \tag{2.27}$$

where the initial conduction phase θ_m can be calculated as

$$\theta_m = arcsin\left(\frac{u_{dc\_min}}{u_{dc\_max}}\right), \theta_m \in \left[0, {}^{\pi}/_{2}\right]. \tag{2.28}$$

It can be easily proved that the grid voltages with different frequencies almost provide the same average inverter output voltage because of the characteristic of the sine function. Generally, because of the voltage and current rating of the inverter and the motor itself, the maximum operation speed of the machine is limited by the voltage and current constraint. The voltage and current constraint can be presented as (2.30), where i_s represents the maximum current and it can be simplified as:

$$\begin{gathered}\left(L_q i_q\right)^2 + \left(L_d i_d + \psi_f\right)^2 \le \frac{u_{dc}^2}{3\omega_r^2} \\ i_d^2 + i_q^2 \le i_s^2\end{gathered} \tag{2.29}$$

for the condition ignoring the voltage drop caused by the stator resistance and assuming the steady-state operation. As for the drive system, the voltage constraint can be presented as:

$$\left(L_q i_q\right)^2 + \left(L_d i_d + \psi_f\right)^2 \leq \frac{u_{dc\_max}^2}{3}\left(\frac{\lambda}{\omega_r}\right)^2. \tag{2.30}$$

It can be seen that the maximum operating speed is proportional to the attenuation coefficient λ and the reduced percentage of the operating speed is consistent with λ. The reduced percentage of the speed can be shown in Fig. 2.15. It can be concluded that the operating speed reduces with the decrease of u_{dcmin}/u_{dcmax} and the minimum percentage is 63.7% when the DC-link voltage decreases to zero.

The constraints and current trajectory with different loads are shown in Fig. 2.16. The solid curves represent the determined voltage constraint and the dotted curves represent the voltage constraint corresponded to the current constraint. The intersection points of the determined voltage constraint and current trajectory represent actual motor operating points. The magnitude of *d*- and *q*-axis current will increase to supply the motor torque for the loads as the *d*-axis current increases. It can be seen that the intersection of the voltage constraint and the current constraint is the motor operating region. As for the reduced DC-link capacitance motor drive system, the voltage constraint will determine the maximum operating speed because of the remarkably reduced DC-link capacitance. The operating points can be obtained by the combination of the determined voltage constraint and the current trajectory with different loads as shown in Fig. 2.16.

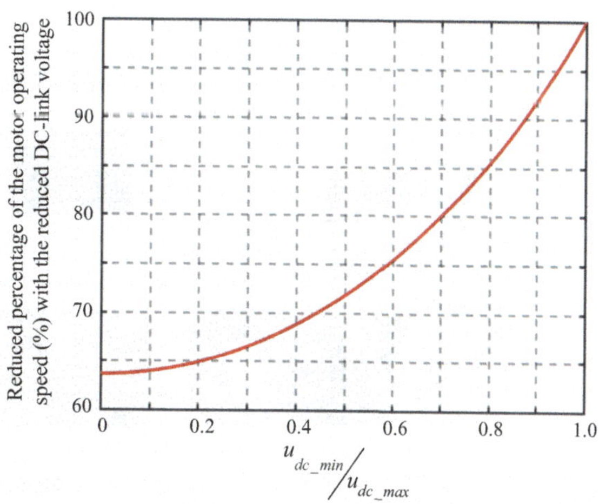

Fig. 2.15 Relationship between the DC-link voltage and the reduced percentage of the motor operating speed

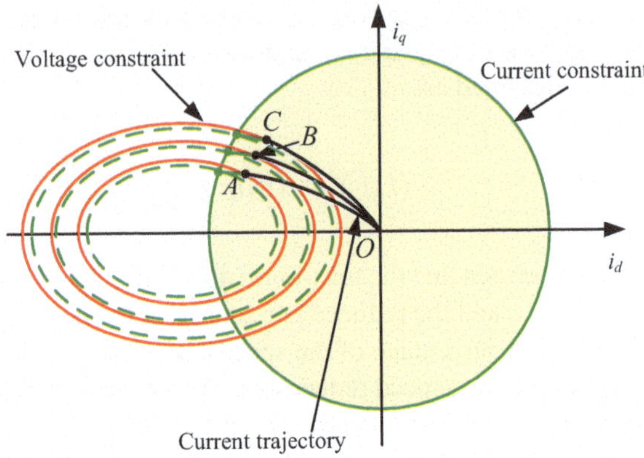

Fig. 2.16 The voltage current constraint of motor operation with different loads

2.5 Experimental Results

For the purpose of demonstrating the effectiveness of the proposed inverter control method, experiments are performed on an air conditioner drive platform without electrolytic capacitors in the DC-link. Figure 2.17 shows the experimental platform, and the parameters of platform are shown in Table 2.2.

The whole control algorithm is realized with a digital signal processor (DSP) TMS320F28034. As the mechanical sensors cannot be installed on the compressor, the motor is operated by the sensor-less control. The bandwidth of current PI controllers is selected as 200 Hz to ensure the stability of the system, the proportional gain K_{PRP}, the resonant gain K_{PRR} and the cut-off frequency ω_{prc} of the PR controller

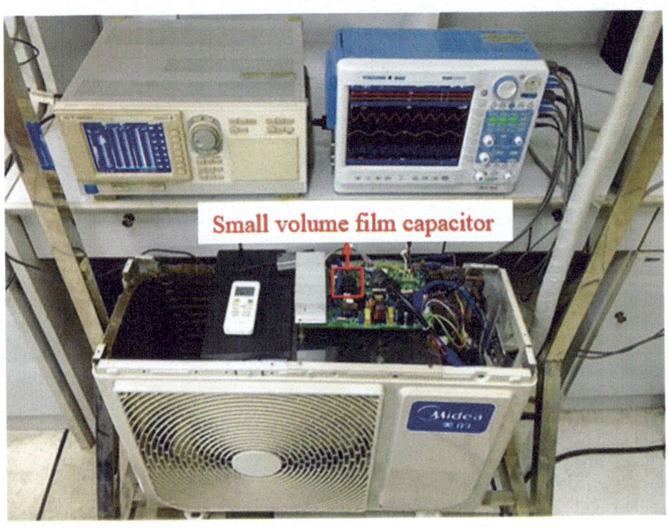

Fig. 2.17 Experimental platform (Air-Conditioner)

Table 2.2 Parameters of the experimental platform

Parameter	Value	Parameter	Value
Grid voltage	220 Vrms	Rated speed	3000 r/min
Grid frequency	50 Hz	Stator resistor	1.48 Ω
Film capacitor	20 μF	*d*-axis inductance	7.9 mH
Line filter inductor	5 mH	*q*-axis inductance	11.7 mH
Switching frequency	10 kHz	Flux linkage	0.11 Wb
Sampling frequency	10 kHz	pole pairs	3
Rated power	1.0 kW		

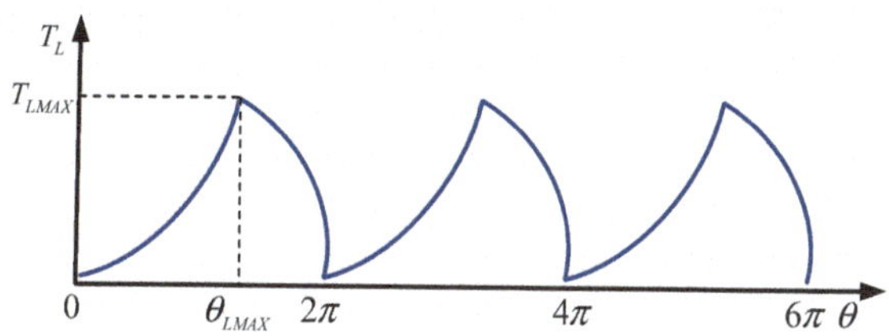

Fig. 2.18 Load torque of the single rotary compressor

are 1.0, 0.1 and 0.02π, respectively. The power meter WT1600 is used to measure the harmonics of the grid current. The mechanical load characteristic of the single rotary IPMSM compressor air conditioner is shown in Fig. 2.18, and it can be seen that the load torque has a large variation per mechanical cycle, which is determined by the operational principle of the single rotary compressor.

Figure 2.19 shows the experimental results with and without the inverter power control loop adopting PR controller and the DC-link voltage control strategy when the compressor operates at 3000r/min. The left waveforms from top to bottom show the DC-link voltage, the actual inverter power, the grid current, and the motor current, respectively.

Meanwhile, the Fourier analysis of the DC-link voltage and grid current are shown in the right part. In Fig. 2.19a, without the inverter power control loop and the DC-link voltage control strategy, it can be seen that the inverter power is not under control and the resonance between the inductor of input filter and the DC-link film capacitor is obvious that cannot be effectively suppressed. The resonant frequency of the drive system is approximately 500 Hz when the line inductor is 5mH and the film capacitor is 20μF. It can be seen that the DC-link voltage fluctuates severely and the harmonics of grid current far exceed the standards of EN61000-3-2, and the harmonics around the resonant frequency are obviously serious. At the same time, the compressor is unable to operate at high speed because the irregular inverter power and

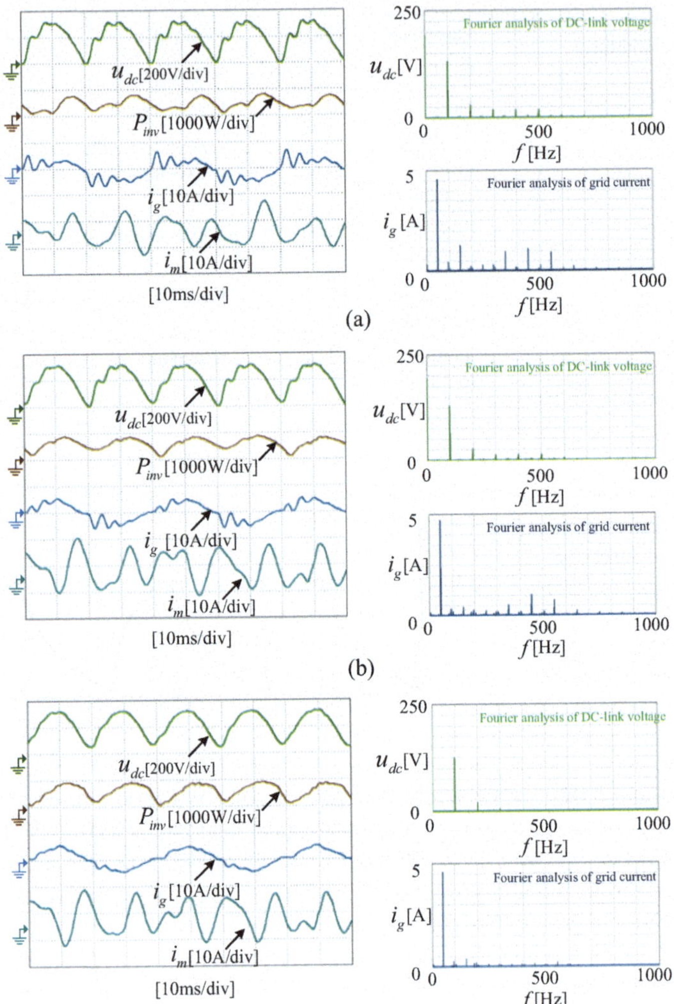

Fig. 2.19 Experimental waveforms when the compressor operates at 3000 r/min. **a** without the inverter power control loop (power factor = 0.891, THD = 51.3%). **b** Only using the inverter power control loop with PR controller (power factor = 0.912, THD = 37.8%). **c** Combining the proposed DC-link voltage control strategy with the inverter power control loop adopting PR controller (power factor = 0.981, THD = 16.0%)

the severe resonance will lead the drive system unstable. In Fig. 2.19b, the inverter power control loop with PR controller is applied and the power factor and THD of the grid side are 0.912 and 37.8%. It can be seen that the grid side performance is improved compared to that in Fig. 2.19a. Combined with the proposed DC-link voltage control strategy, the inverter power control loop with PR controller is applied in Fig. 2.19c. It can be seen that the grid side performance is obviously improved and also the harmonics of grid current can be obviously suppressed. The power factor and THD of the grid side are 0.981 and 16.0%. It can be concluded from Fig. 2.19 that only using the inverter power control loop is difficult to effectively control the inverter power and suppress the harmonics of the grid current. The fluctuation of the DC-link voltage is not under control, which reduces the conduction angle of the diode rectifier and limits the improvement of the power factor. The resonance can be suppressed whereas the performance is limited. Based on the inverter power control loop, better performance can be achieved by applying the proposed DC-link voltage control, which can also shape the DC-link voltage effectively despite the large load variation of the compressor. Meanwhile, the resonance can be effectively suppressed for the enhanced inverter power control performance. As a result, it can be seen from the Fourier analysis that the harmonics of DC-link voltage and grid current can be obviously suppressed.

In order to compare the control performance of PR controller and repetitive controller, in Fig. 2.20a, the repetitive controller is applied in the inverter power control loop and the power factor and THD of the grid side are 0.903 and 38.9%. The corresponding power factor and THD of the PR controller are 0.912 and 37.8%. Combined with the proposed DC-link voltage regulation strategy, the inverter power

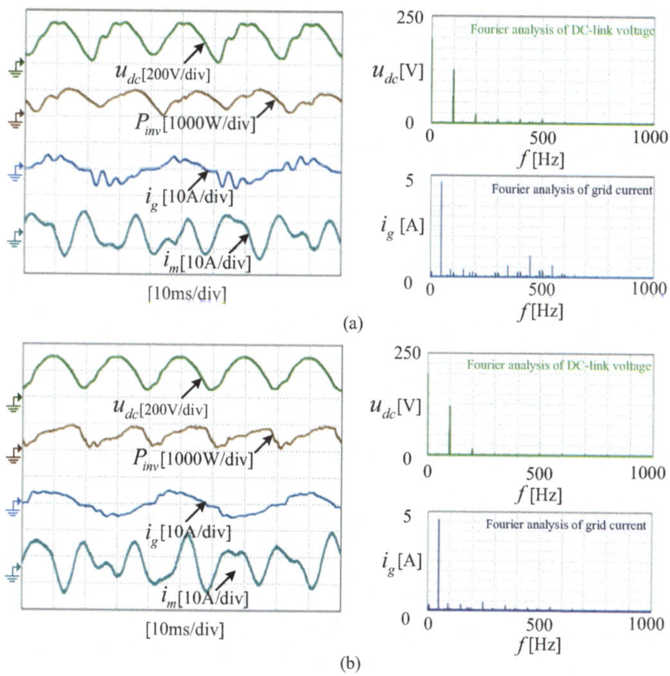

Fig. 2.20 Experimental waveforms adopting repetitive controller when the compressor operates at 3000 r/min. **a** Only using the inverter power control loop with repetitive controller (power factor = 0.903, THD = 38.9%). **b** Combining the proposed DC-link voltage regulation strategy with the inverter power control loop adopting repetitive controller (power factor = 0.973, THD = 17.5%)

control loop with repetitive controller is applied in Fig. 2.20b. The power factor and THD of the grid side are 0.973 and 17.5% and the corresponding power factor and THD of the PR controller are 0.981 and 16.0%. It can be seen that the control performance of PR controller is better than the applied repetitive controller, because the PR controller is more effective to the sinusoidal signals than the repetitive controller as described in the previous section.

Figure 2.21 shows the experimental results applying the inverter power control loop with PR controller and the proposed DC-link voltage regulation strategy when the compressor operates at 5000r/min. It can be seen that the control performance becomes better as the compressor speed increases. The motor current becomes more regular than that of the motor operates at 3000r/min, which means that the effect of the load torque fluctuation on the drive system is reduced and it can benefit the control performance.

In order to compare the efficiency, the grid input power is measured by a power meter WT1600, which is calculated by the root mean square (RMS) values of the grid voltage and current. The measured values are 569 W, 571 W and 573 W by applying the conventional methods (in Fig. 2.19a) and the proposed method (in Fig. 2.19b and

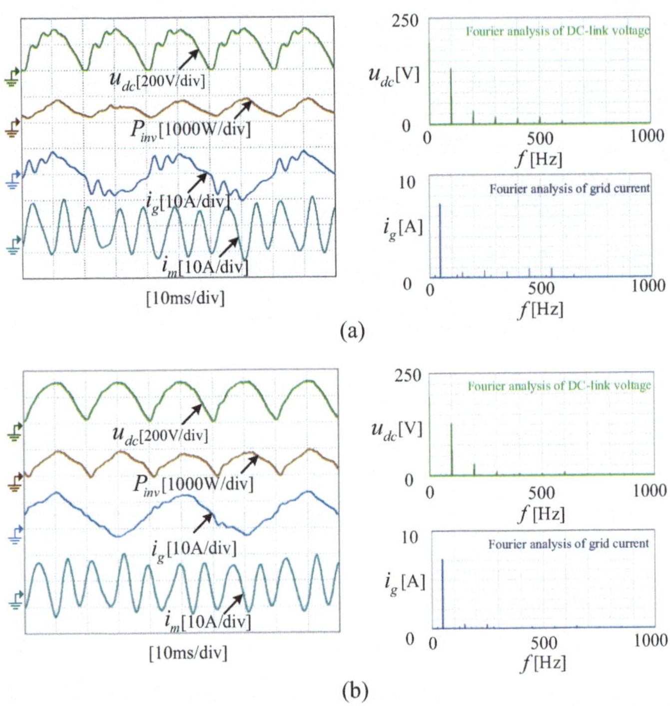

Fig. 2.21 Experimental waveforms when the compressor operates at 5000 r/min. **a** Only using the inverter power control loop with PR controller (power factor = 0.964, THD = 26.3%). **b** Combining the proposed DC-link voltage regulation strategy with the inverter power control loop adopting PR controller (power factor = 0.991, THD = 10.7%)

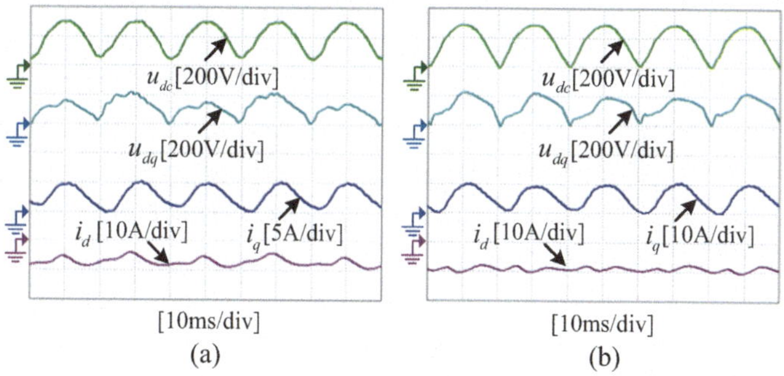

Fig. 2.22 Experimental waveforms of the magnitude of voltage vector, *d*- and *q*- axis current components using the proposed inverter power control method. **a** 3000 r/min. **b** 5000 r/min

c), respectively. It can be seen that the efficiency of the proposed method is slightly lower because of the fluctuated motor power, which can motivate high-order power ripple. As a result, the power ripple will increase the harmonic iron loss compared to the drive system that outputs constant motor power. Meanwhile, the high-order power ripple will increase the eddy current loss, which can also reduce the motor drive efficiency. It can be concluded that the desired fluctuated inverter power can benefit the grid side quality, whereas the caused power ripple will slightly reduce the drive system efficiency and it can be compensated by the optimized design of the motor.

Figure 2.22 shows the voltage and current waveforms when the compressor operates at 3000 and 5000 r/min. The waveforms from top to bottom are the DC-link voltage, the magnitude of the voltage vector, and the *d*- and *q*- axis current components, respectively. The magnitude of the voltage vector is lower than the DC-link voltage, which guarantees the stable operation of the electrolytic capacitor-less compressor drive system. The *q*-axis current fluctuates with the DC-link voltage. Although the tracking performance of the *d*-*q* axis currents is limited by the bandwidth of the current PI controllers, the inverter power of the reduced DC-link capacitance motor drive system can be effectively controlled. As a result, the harmonics of the grid current can satisfy the standards of EN61000-3-2, and the detailed harmonics measured by WT1600 are shown in Fig. 2.23. It can be seen that the harmonics are far lower than the required values of the standard.

Figure 2.24 shows the grid voltage and current waveforms when the compressor operates at 3000 and 5000 r/min. The harmonic analysis of the current waveforms has been operated based on a longer time interval. It can be seen that the grid current is approximate to be the sinusoidal waveform and the harmonics within entire frequency range can greatly satisfy the standards of EN61000-3-2. The experimental results of the power factor and the THD at different operating speeds are shown in Fig. 2.25. The proposed inverter power control method can effectively maintain high power factor and low THD in the wide speed range. The maximum power factor is 0.992. The harmonics can satisfy EN61000-3-2 and the minimum THD is 10.2%. The performance of the electrolytic capacitor-less drive system can meet the requirement of the

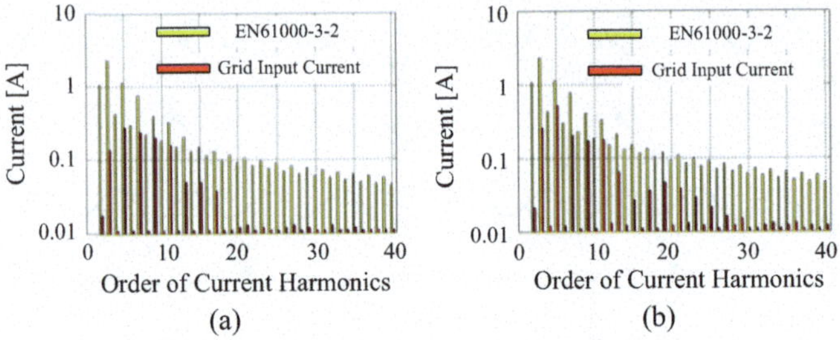

Fig. 2.23 Comparison with the EN61000-3-2 limits and the grid current harmonics applying the proposed inverter power control method. **a** 3000 r/min. **b** 5000 r/min

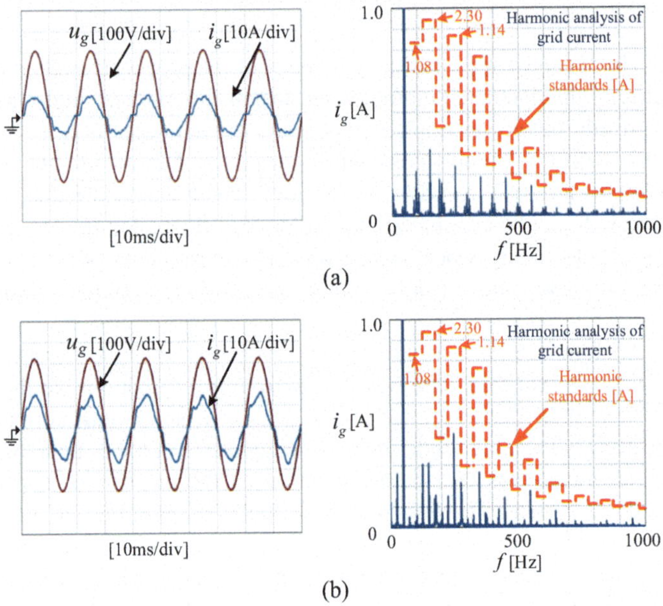

Fig. 2.24 Experimental waveforms of the grid voltage and grid current and the harmonic analysis of grid current. **a** 3000 r/min. **b** 5000 r/min

air conditioner applications. It benefits the cost reduction and reliability improvement of the system by using the proposed control method.

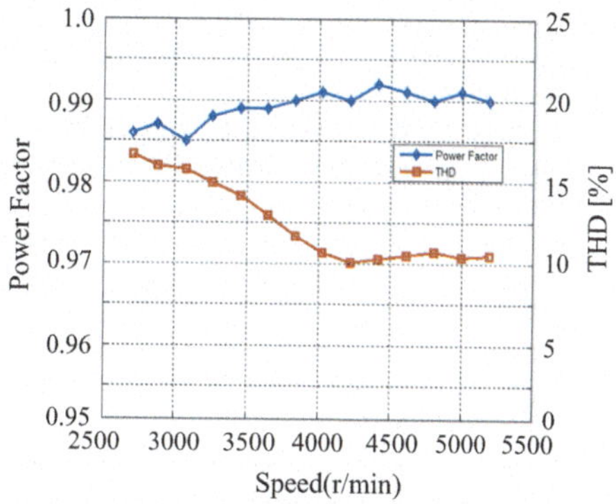

Fig. 2.25 Power factor and THD in the overall speed range of the compressor

2.6 Summary

This chapter presents an inverter power control strategy based on the DC-link voltage control applied in the reduced DC-link capacitance drive system for PMSM. For the reduced capacity of the DC-link capacitor and the elimination of the PFC circuit, the inverter power closely affects the power factor and the THD of the grid side. The inverter power control loop with PR controller is established to regulate the inverter power, which can achieve a high gain at the fluctuation frequency of the DC-link voltage. Besides, the parameters of PR controller are easy to design. Only applying the inverter power control loop, the actual inverter power is difficult to totally track the inverter power reference because of the low bandwidth of the PI current controllers and the power coupled within the *d-q* reference frame. Therefore, a power compensation method based on the DC-link voltage control is applied to diminish the inverter power error, which does not depend on the precise calculation and is easy to realize. By using the proposed inverter power control strategy based on the DC-link voltage regulation, the power factor of the drive system can reach 0.992, and the THD of the grid current can greatly satisfy the regulations of the EN61000-3-2 without the PFC circuit. The effectiveness of the proposed method is verified by experimental results.

References

1. J.Y. Lee, An EL capacitorless EV on-board charger using harmonic modulation technique. IEEE Trans. Ind. Electron. **61**(4), 1784–1787 (2014)
2. J.C.W. Lam, P.K. Jain, A high power factor, electrolytic capacitor-less AC-input LED driver topology with high frequency pulsating output current. IEEE Trans. Power Electron. **30**(2), 943–955 (2015)
3. Y. Yang, X. Ruan, L. Zhang, J. He, Z. Ye, Feed-forward scheme for an electrolytic capacitor-less AC/DC LED driver to reduce output current ripple. IEEE Trans. Power Electron. **29**(10), 5508–5517 (2014)

4. Y. Shi, R. Li, Y. Xue, H. Li, High-frequency-link-based grid-tied PV system with small DC link capacitor and low-frequency ripple-free maximum power point tracking. IEEE Trans. Power Electron. **31**(1), 328–339 (2016)
5. N. Zhao, G. Wang, D. Xu, L. Zhu, G. Zhang, J. Huo, Inverter power control based on DC-link voltage regulation for IPMSM drives without electrolytic capacitors. IEEE Trans. Power Electron. **33**(1), 558–571 (2018)
6. K. Inazuma, H. Utsugi, K. Ohishi, H. Haga, High power factor single-phase diode rectifier driven by repetitively controlled IPM motor. IEEE Trans. Ind. Electron. **60**(10), 4427–4437 (2013)
7. Y. Son, J.I. Ha, Direct power control of three phase inverter for grid input current shaping of single phase diode rectifier with small DC-link capacitor. IEEE Trans. Power Electron. **30**(7), 3794–3803 (2015)
8. P. Magne, D. Marx, B. Nahid-Mobarakeh, S. Pierfederici, Large-signal stabilization of a DC-link supplying a constant power load using a virtual capacitor: impact on the domain of attraction. IEEE Trans. Ind. Appl. **48**(3), 878–887 (2012)
9. A. Kuperman, Proportional-resonant current controllers design based on desired transient performance. IEEE Trans. Power Electron. **30**(10), 5341–5345 (2015)
10. D.G. Holmes, T.A. Lipo, B.P. McGrath, W.Y. Kong, Optimized design of stationary frame three phase AC current regulators. IEEE Trans. Power Electron. **24**(11), 2417–2426 (2009)
11. T. Ye, N. Dai, C. Lam, M. Wong, J. Guerrero, Analysis, design and implementation of a quasi-proportional-resonant controller for a multi-functional capacitive-coupling grid-connected inverter. IEEE Trans. Ind. Appl **52**(5),4269–4280 (2016)

Chapter 3
Resonance Suppression Between Line Inductor and DC-Link Capacitor

The harmonic of the grid current is an important issue in the reduced DC-link capacitance motor drive system. An input inductor has been applied in the grid side to improve the grid current quality. Whereas the LC resonance between the line inductor and the DC-link film capacitor leads to the additional harmonics at the resonant frequency [1–4]. It will stimulate the grid current distortion and pollute the power grid. Meanwhile, the drive system behaving the negative impedance characteristic is the constant power loads (CPLs) system. As the DC-link voltage increases/decreases, the DC-link current will decrease/increase, which is contrary to the DC-link voltage. The interaction between the negative impedance characteristic and the LC resonance may cause the instability of the drive system [5–7]. The general solution is to increase the system damping to suppress the resonance phenomenon by applying passive damping or active damping methods. As for the actual application, active damping methods are preferred considering the power loss, the system size and the cost [8–10].

3.1 Analysis of LC Resonance

3.1.1 Drive System Model Construction

As shown in Fig. 2.1, the grid current i_g can be calculated as

$$i_g = i_{inv} + i_c \tag{3.1}$$

The equivalent source voltage u_g can be calculated as

$$u_g = L_g \frac{di_g}{dt} + R_g i_g + u_{dc} \tag{3.2}$$

© The Author(s), under exclusive license to Springer Nature Singapore Pte Ltd. 2021

G. Wang et al., *Reduced DC-link Capacitance AC Motor Drives*,
https://doi.org/10.1007/978-981-15-8566-1_3

where L_g and R_g are the line inductance and the equivalent resistance, respectively.

Actually, the equivalent current source i_{inv} can be presented by

$$i_{inv} = \frac{P_L}{u_{dc}} = \frac{P_L}{u_{dc,0} + \tilde{u}_{dc}} \tag{3.3}$$

where P_L and $\tilde{u}_{dc}$ are the load of the motor and the deviation of the DC-link voltage, respectively. The linearized equation of (3.3) can be presented as

$$i_{inv} = \frac{P_L}{u_{dc,0}} - \frac{P_L}{u_{dc,0}^2}\tilde{u}_{dc}. \tag{3.4}$$

Meanwhile, the current of DC-link capacitor i_c meets

$$i_c = C_{dc}\frac{du_{dc}}{dt} \tag{3.5}$$

Combining (3.1), (3.2), (3.4) and (3.5), the small signal equivalent model of the drive system can be obtained and the characteristic polynomial of the drive system can be presented as

$$s^2 + (\frac{R_g}{L_g} - \frac{P_L}{C_{dc}u_{dc,0}^2})s + \frac{1}{L_g C_{dc}}(1 - \frac{P_L R_g}{u_{dc,0}^2}) = 0. \tag{3.6}$$

3.1.2 *Stability Analysis of Drive System*

According to the Routh criterion, the stability conditions of the drive system can be obtained as follows:

$$\begin{cases} 1 - \frac{P_L R_g}{u_{dc,0}^2} > 0 \\ \frac{R_g}{L_g} - \frac{P_L}{C_{dc}u_{dc,0}^2} > 0. \end{cases} \tag{3.7}$$

The negative input-impedance shown in (3.4) leads to the negative parts of (3.7), which causes the instability as load power P_L increases. Assuming the input-impedance is positive, the drive system will be stable regardless of P_L. As a typical constant load power system, the motor drive system is deeply affected by the characteristic of the negative input-impedance. The stability issue will be more serious as the DC-link capacitance reduces. Generally, the mean value of the equivalent voltage source meets

$$u_{dc,0}^2 \gg P_L R_g \tag{3.8}$$

$$\frac{C_{dc}}{P_L} > \frac{L_g}{R_g u_{dc,0}^2}. \tag{3.9}$$

3.1.3 Influence of DC-Link Capacitance on Drive System

According to the above analysis, it can be seen that the stability of the drive system is related to the whole components of the drive circuit. If the conventional DC-link electrolytic capacitor is replaced by the film capacitor, the capacity will be remarkably decreased. The pole distribution as C_{dc} reduces is shown in Fig. 3.1. It can be seen that poles of the drive system moves from the left half plane to the right half plane as C_{dc} reduces, and the drive system becomes unstable when poles reach the right half plane.

Hence the stability of the drive system will be reduced. Meanwhile, the DC-link voltage and the grid current will be resonant due to the line inductor and the reduced DC-link capacitor. The resonant frequency can be calculated as:

$$\omega_{res} = \frac{1}{\sqrt{L_g C_{dc}}}. \tag{3.10}$$

Harmonics of the grid current caused by the resonance will exceed the standard requirements of EN61000-3-2 easily. Therefore, it is necessary to improve the

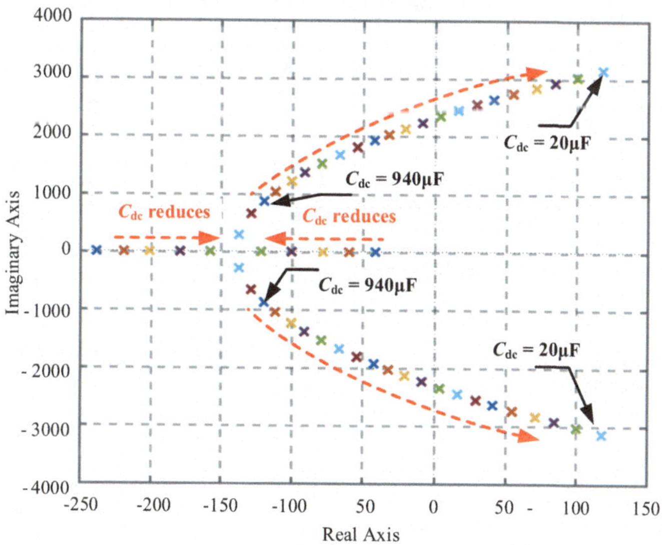

Fig. 3.1 Pole distribution of the drive system as C_{dc} reduces

stability of the drive system and suppress the resonant DC-link voltage and the grid current. A stabilization method based on the direct damping current will be introduced in the next section.

3.2 DC-Link Voltage Feedback Based Active Damping Control Method

3.2.1 *Principle of Active Damping Control*

The instability of the reduced DC-link capacitance motor drive system is caused by the unexpected characteristic equation as presented in the previous discussion. It is necessary to add an active damping part to reconstruct the poles of characteristic equation for the drive system. As shown in Fig. 3.2a, the active damping part can be regarded as a damping current source i_{damp} parallel to the inverter. Hence the solution is to superpose the direct damping current upon the inverter current. However, it is not easy to regulate the current flowing from DC-link to the inverter in the reduced DC-link capacitance motor drive system. Meanwhile, the resonant frequency ω_{res} is usually higher than the bandwidth of the current controller. As a result, it is difficult to directly generate the direct damping current. Fortunately, the effect of the direct damping current can be achieved by the damping power P_{damp} shown in Fig. 3.2b. The generation of P_{damp} can be directly realized by the voltage command instead

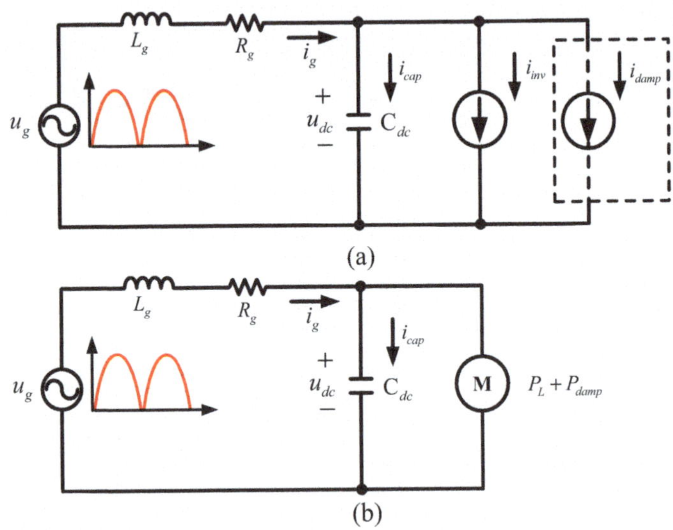

Fig. 3.2 Drive system with damping method. **a** Stabilizing the drive system with direct damping current. **b** Equivalent damping power

of the current command, which can overcome the delay effect of the relatively low bandwidth current controller in the reduced DC-link capacitance drive system.

3.2.2 *Direct Damping Current to Stabilize Drive System*

As the PFC circuit is eliminated in the reduced DC-link capacitance motor drive system, the resonance between the line inductor and the film capacitor will affect the DC-link voltage. Hence the resonance can be detected by the DC-link voltage and i_{damp} is achieved by the operation of the DC-link voltage. Applying three different basic control laws, i.e. the proportional, integral and differential, the grid current i_g can be presented as:

$$i_g = i_{inv} + i_c + K_{dampP}\tilde{u}_{dc}, \text{ P - type} \tag{3.11}$$

$$i_g = i_{inv} + i_c + K_{dampI}\int \tilde{u}_{dc}dt, \text{ I - type} \tag{3.12}$$

$$i_g = i_{inv} + i_c + K_{dampD}\frac{d\tilde{u}_{dc}}{dt}, \text{ D - type} \tag{3.13}$$

where K_{dampP}, K_{dampI}, and K_{dampD} are gains of control laws, which are the proportional, integral and differential, respectively. The drive system damped by i_{damp} is shown in Fig. 3.3. It can be seen obviously that i_{damp} has an influence on the current of DC-link capacitor i_c, which can shape the DC-link voltage to suppress the resonance.

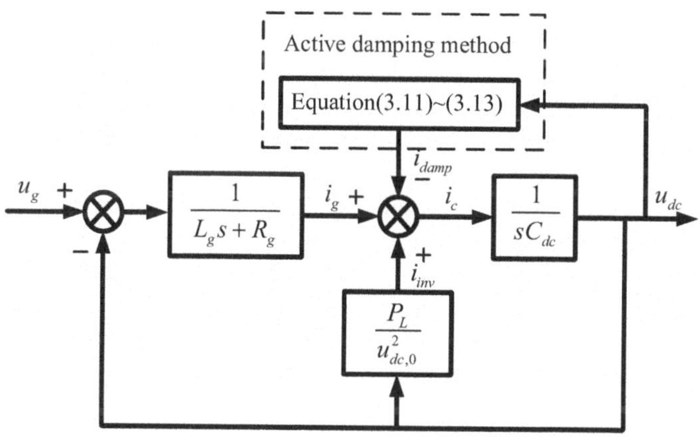

Fig. 3.3 Block diagram of the drive system with i_{damp}

3.2.3 Stability Analysis Using Routh-Hurwitz Criterion

As shown in Fig. 3.4, the transfer function of the equivalent model with different control laws can be obtained and the new characteristic polynomial could be evaluated by the Routh-Hurwitz criterion. The typical third-order characteristic polynomial in the Laplace variable s is represented as:

$$a_0 s^3 + a_1 s^2 + a_2 s + a_3 = 0 \tag{3.14}$$

where a_0, a_1, a_2 and a_3 are coefficients of the polynomial. The stability criterion is that all the coefficients are positive and the relationship should meet:

$$a_1 a_2 > a_0 a_3. \tag{3.15}$$

Hence the stability criterions with different control laws of DC-link voltage could be obtained by the Routh-Hurwitz criterion. The stability constraints of K_{dampP}, K_{dampI}, and K_{dampD} are inferred as follows:

$$K_{dampP} > Max\left(\frac{R_g P_L - u_{dc,0}^2}{R_g u_{dc,0}^2}, \frac{L_g P_L - R_g C_{dc} u_{dc,0}^2}{L_g u_{dc,0}^2}\right) \tag{3.16}$$

$$K_{dampI} < \frac{\left(u_{dc,0}^2 - R_g P_L\right)\left(R_g C_{dc} u_{dc,0}^2 - L_g P_L\right)}{L_g^2 P_L u_{dc,0}^2} \tag{3.17}$$

$$K_{dampD} > \frac{L_g P_L - R_g C_{dc} u_{dc,0}^2}{R_g u_{dc,0}^2}. \tag{3.18}$$

It should be noticed that stability criterions of differential and integral control laws also have other constant constraints except the stability constraint of K_{dampI} and K_{dampD} as shown in (3.17) and (3.18). In order to evaluate the drive system stability directly, the pole distributions with different control laws are depicted in Fig. 3.4.

It can be seen that pole distribution is adjusted as the values of K_{dampP}, K_{dampI} and K_{dampD} increase and both the proportional and differential operation of the DC-link voltage can stabilize the drive system effectively. However, the integral operation will increase the system order and lead to the instability. Compared with the differential operation, the proportional operation of the DC-link voltage can achieve better performance and easier realization, because the differential operation is sensitive to the noise of the drive system. Meanwhile, once combining the proportional operation with the differential operation or the integral operation, the performance is not better and parameters are difficult to be optimized. Hence the proportional operation of the DC-link voltage is applied to generate the direct damping current in the proposed method.

Comparing (3.16) with (3.7), It can be seen that numerators of (3.16) are equivalent to the ones in (3.7). It proves the stability condition for K_{dampP} closely related to the

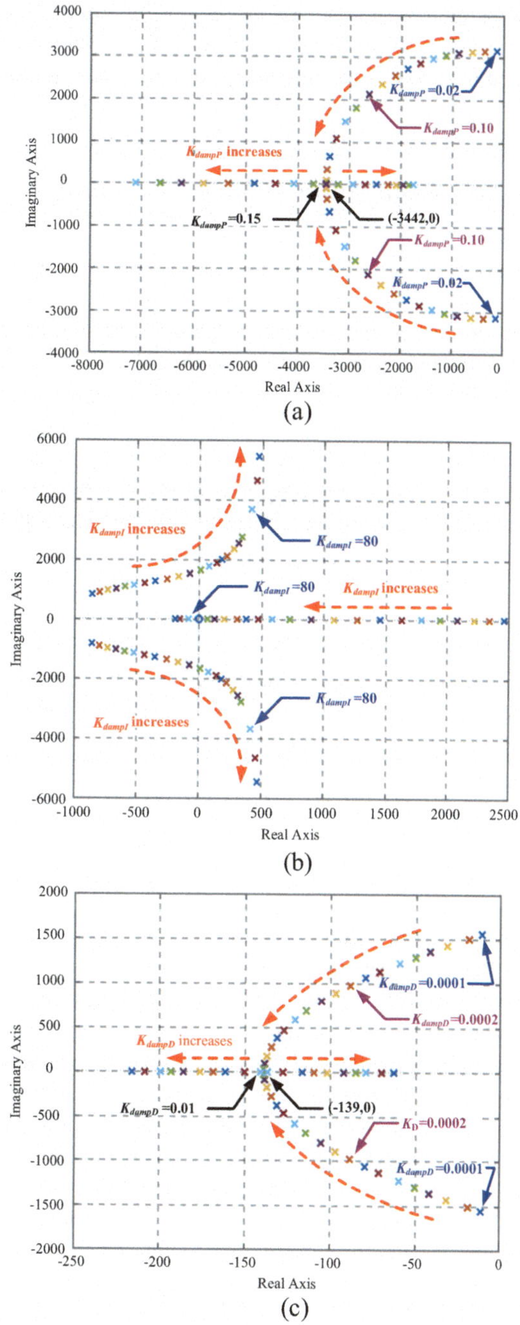

Fig. 3.4 Pole distribution of the drive system with different control laws.
a Proportional operation of the DC-link voltage.
b Integral operation of the DC-link voltage.
c Differential operation of the DC-link voltage

undamped system characteristics. The Routh-Hurwitz criterion applied in (3.16) determines the minimum value of K_{dampP} when the system becomes stable. It can be seen from Fig. 3.4a that the system becomes stable when K_{dampP} is larger than 0.02.

3.2.4 Realization of Direct Damping Current

The direct damping current is parallel to the DC-link capacitor and it is difficult to be accurately generated and controlled. The solution to realize the direct damping current can be transformed to generate the equivalent damping power P_{damp}, which is generated by the inverter power and it can be calculated as follows:

$$P_{damp} = u_{dc} i_{damp}. \tag{3.19}$$

Ignoring the switching loss of the inverter, the motor power can be regarded as the inverter power, which can be presented as follows:

$$P_{inv} = 1.5\left(u_\alpha i_\alpha + u_\beta i_\beta\right). \tag{3.20}$$

The generation of the damping power can be achieved by the additional current or voltage command. Compared with the current command, the solution based on the voltage command can overcome the limited bandwidth of the current controller and accurately generate the damping power based on the inverter power, which can be presented as

$$P_{inv} + P_{damp} = 1.5\left[(u_\alpha + \Delta u_\alpha)i_\alpha + \left(u_\beta + \Delta u_\beta\right)i_\beta\right] \tag{3.21}$$

where Δu_α and Δu_β are voltage commands for modification in the stationary frame.

The overall block diagram of the active damping control method is depicted in Fig. 3.5. Generally, the outputs of current controllers will be decoupled by the feedforward voltage and transformed by the Park transformation as the stationary voltage reference vector $\boldsymbol{u}_{\alpha\beta}$. The damping current i_{damp} can be generated by the DC-link voltage, which is processed by a high pass filter (HPF) and the proportional operation. Meanwhile, P_{damp} can be generated by the multiplication of i_{damp} and u_{dc}, which is based on (3.19), and the voltage command Δu_α and Δu_β should be added to $\boldsymbol{u}_{\alpha\beta}$. However, the voltage command can also be regarded as a disturbance to the motor drive system. Hence the distribution of Δu_α and Δu_β as presented in (3.21) is an important issue. In order to solve it, the minimum effect of the voltage command on the drive system should be considered in the voltage command generator and the magnitude of the voltage command vector should be as small as possible.

The damping power can be presented as follows:

$$P_{damp} = \Delta \boldsymbol{u}_{\alpha\beta} \cdot \boldsymbol{i}_{\alpha\beta} \tag{3.22}$$

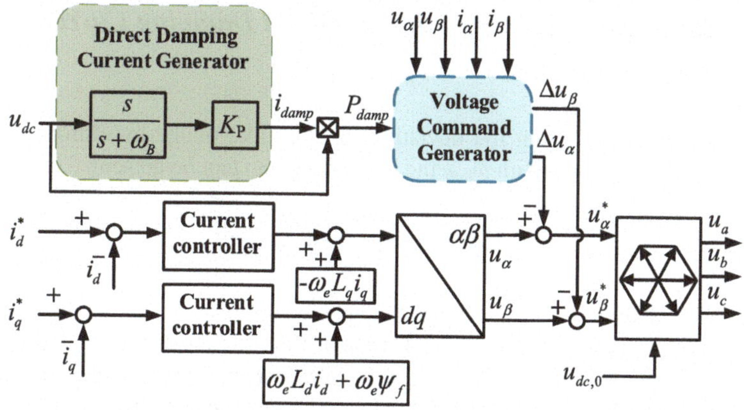

Fig. 3.5 Overall diagram of the active damping control method

where $\Delta \boldsymbol{u}_{\alpha\beta}$ and $\boldsymbol{i}_{\alpha\beta}$ are the additional voltage command and current vector in the stationary frame, respectively. As shown in Fig. 3.6, the possible direction of the voltage command is infinite, all of which could generate the same damping power P_{damp}. The minimum magnitude of voltage command vector is located in the direction which is parallel to the current vector, which can reduce the impact of the voltage commands on the drive system. Hence voltage commands of the active damping method are generated based on the damping power P_{damp}.

Fig. 3.6 Realization of voltage command generator

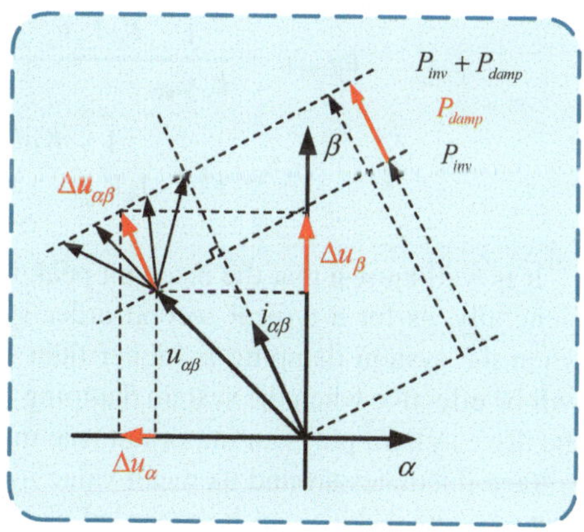

3.2.5 Parameters Determination of Direct Damping Current Generator

A HPF is applied to obtain the deviation of DC-link voltage by eliminating the fundamental component. With ω_{dcB} the bandwidth, $\tilde{u}_{dc}$ can be presented as

$$\tilde{u}_{dc} = \frac{s}{s + \omega_{dcB}} u_{dc}. \tag{3.23}$$

The determination of ω_{dcB} mainly depends on the resonant frequency that can be calculated according to (3.10), and it should be low enough to guarantee that harmonics of the DC-link voltage can be totally detected. Meanwhile, the excessive low bandwidth will extract the fundamental component. Hence the HPF cutoff frequency is set as 250 Hz as a compromise. The magnitude attenuation around the resonant frequency is about 0.9, which can effectively extract the resonant component of the DC-link voltage. The magnitude attenuation is an important concern of the bandwidth determination of the HPF.

The closed loop transfer function after adopting the damping method can be presented as:

$$G_{dampb}(s) = \frac{1}{s^2 + b_{damp0}s + b_{damp1}} \tag{3.24}$$

where

$$\begin{aligned} b_{damp0} &= \frac{1}{L_g C_{dc}} \frac{K_P L_g u_{dc,0}^2 + R_g C_{dc} u_{dc,0}^2 - L_g P_L}{u_{dc,0}^2}, \\ b_{damp1} &= \frac{1}{L_g C_{dc}} \frac{K_P R_g u_{dc,0}^2 + u_{dc,0}^2 - R_g P_L}{u_{dc,0}^2}. \end{aligned} \tag{3.25}$$

It is well known that the resonant peak is closely related to the system damping. Generally, as for a typical second-order system, the resonant peak will disappear when the system damping is higher than 0.707. Hence the damping performance will be effective when the system damping reaches 0.707. Actually, the overshoot of the drive system plays an important role in the dynamic performance. The DC-link voltage fluctuates around its mean value in the reduced DC-link capacitance motor drive system, which can be regarded as a dynamic process. As K_P increases, the drive system becomes a critical damping system, and the damping performance will be improved when the overshoot is completely eliminated. The value of K_P for the critical damping of the drive system is presented as:

$$b_{damp0}^2 - 4b_{damp1} = 0. \tag{3.26}$$

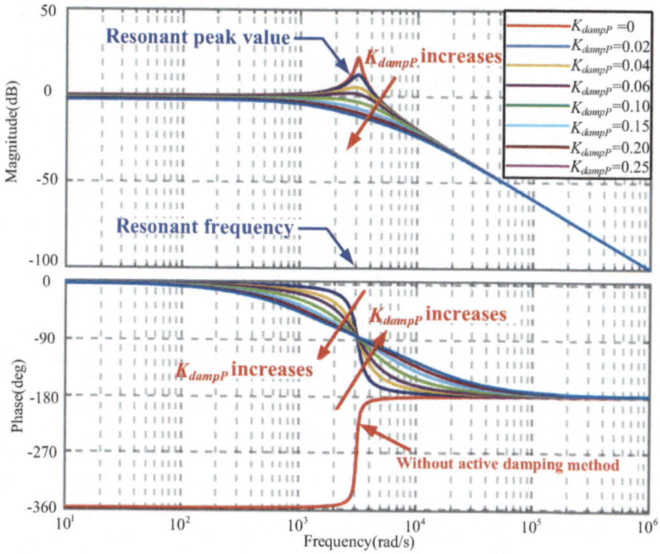

Fig. 3.7 Bode diagrams of the drive system as K_P increases

Bode diagrams with different values of K_{dampP} are shown in Fig. 3.7. It can be seen that the resonant peak value is damped gradually as K_{dampP} increases and the proper value of K_{dampP} occurs when the system becomes a critical damping system, which is consistent with the breakaway point of the pole distribution shown Fig. 3.4a.

3.2.6 Experimental Results

Figure 3.8 shows the system performance comparison without and with the active damping control method when the motor operates at 3000 rpm. The waveforms in left from top to bottom are the grid voltage, the DC-link voltage, the grid current, and the motor current, respectively. The waveforms in right are the Fourier analysis of the DC-link voltage and the grid current.

It can be seen from Fig. 3.8a that the resonance between the line inductor and the DC-link capacitor leads to the significant distortions of the DC-link voltage and grid current, which are consistent with the simulation results. Harmonics around the resonant frequency exceed the harmonics standards of EN61000-3-2, which pollute the power grid and reduce the stability of the drive system. In Fig. 3.8b and c, the proposed active damping control method has been applied with different values of K_{dampP} to suppress the resonance and enhance the system stability. It can be seen that the suppression performance of the resonant grid current and DC-link voltage is obvious when K_{dampP} is 0.05. Harmonics of the grid current satisfy the standards of EN61000-3-2. The performance can be optimized as shown in Fig. 3.8c when K_{dampP} is 0.15. It can be seen from Fig. 3.8 that the peak value of the resonant frequency is high

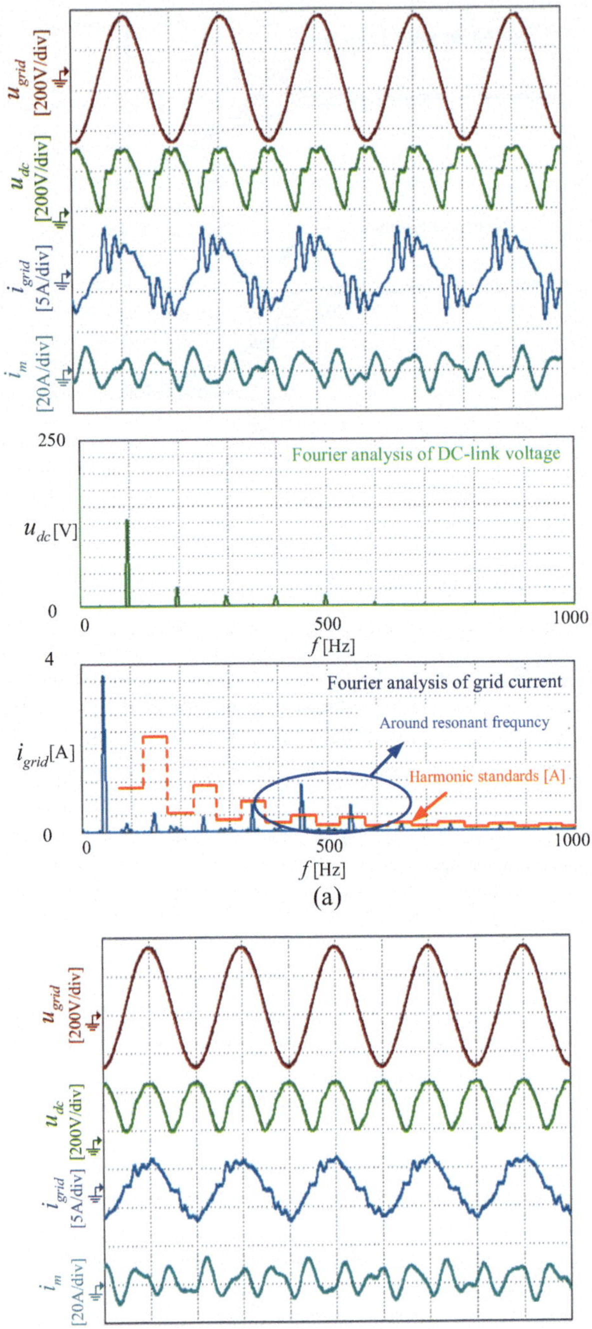

Fig. 3.8 Experimental waveforms of the grid voltage, DC-link voltage, grid current and motor current. **a** Without the proposed method. **b** With the proposed method, $K_{dampP} = 0.05$. **c** With the proposed method, $K_{dampP} = 0.15$

Fig. 3.8 (continued)

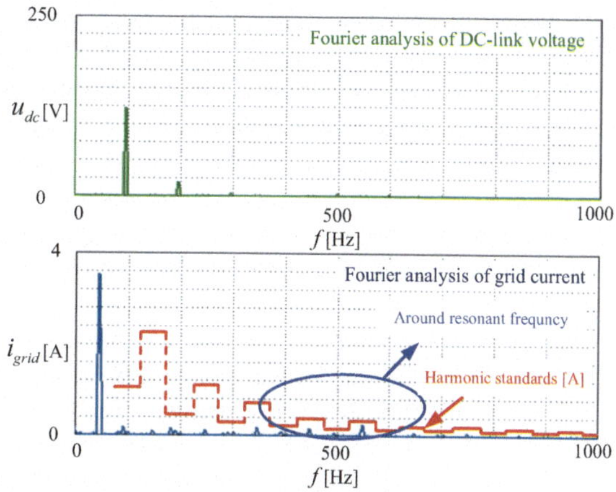

(b)

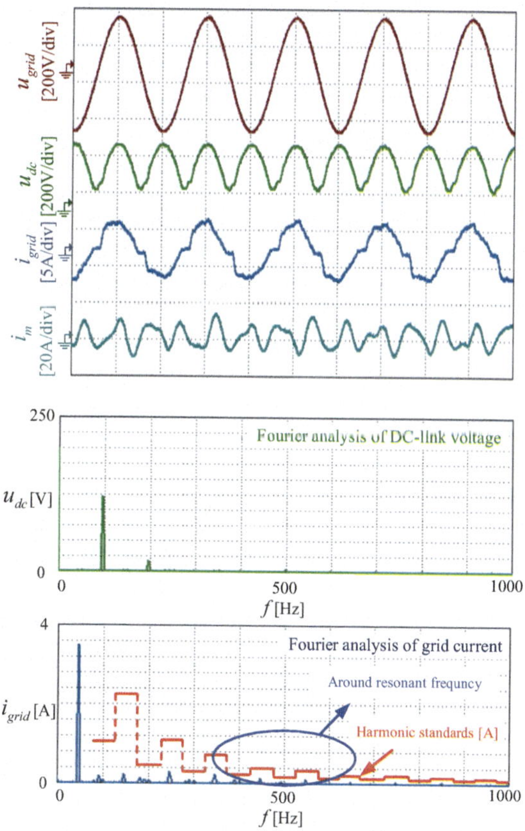

(c)

without the proposed active damping method. As a result, harmonics around resonant frequency are severe in Fig. 3.8a. The effect of the active damping method mainly focuses on the frequency domain around the resonant frequency. Hence magnitudes of harmonics around the resonant frequency are reduced in Fig. 3.8b than the ones in Fig. 3.8a, which are reduced further as K_{dampP} increases in Fig. 3.8c. It means that the performance can be improved by adjusting the direct damping current to the drive system. Meanwhile, according to the previous parameter determination, the optimized value of K_{dampP} is set as 0.15, which is consistent with breakaway point of the pole distribution shown in Fig. 3.4a.

The stationary frame voltage and current waveforms with the proposed method are shown in Fig. 3.9a. Waveforms from top to bottom are the stationary frame voltage $u_{\alpha\beta}$ and current $i_{\alpha\beta}$, respectively. The motor current is irregular due to the fluctuated DC-link voltage and motor load, which is also periodic fluctuated with the rotor mechanical angle. The motor torque with the active damping method is

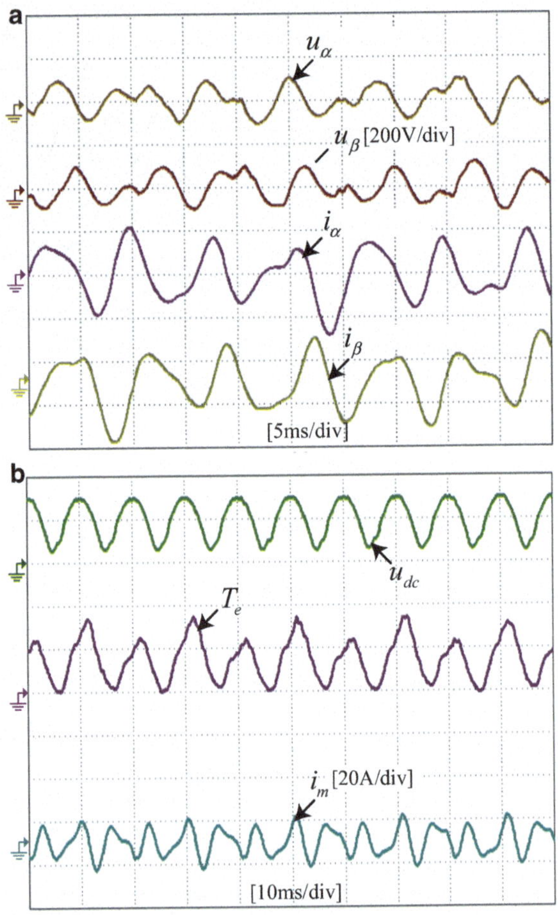

Fig. 3.9 Experimental waveforms with the proposed method, K_P= 0.15. **a** The stationary frame voltage and current. **b** The DC-link voltage, motor torque and motor current

shown in Fig. 3.9b. Experimental waveforms from top to bottom are the DC-link voltage, the motor torque, and the motor current, respectively. It can be seen that the instantaneous torque is fluctuated, and the average value is determined by the fluctuated load torque.

In order to evaluate the effect of voltage commands on the motor current, the Fourier analysis of the motor current without and with the proposed active damping method is shown in Fig. 3.10. It can be seen that harmonics around the resonant frequency (500 Hz) without and with the active damping method are almost the same. Hence the grid harmonic current is suppressed by stabilizing the drive system. Harmonics of the grid current (around the resonant frequency) are not transferred to the motor control. The effect of voltage commands on the motor current around the resonant frequency is critically limited, because the impedance of the motor is high around the resonant frequency. The aim of the active damping method is to stabilize the drive system and damp the resonance by generating the inverter damping power P_{damp}, which is closely related to the instantaneous inverter power. The average inverter power is determined by the motor load, which is closely related to the effective value of the motor current. Hence the fundamental motor current is almost the same and harmonics around the fundamental frequency change slightly for the modified transient inverter power, which are affected by voltage commands.

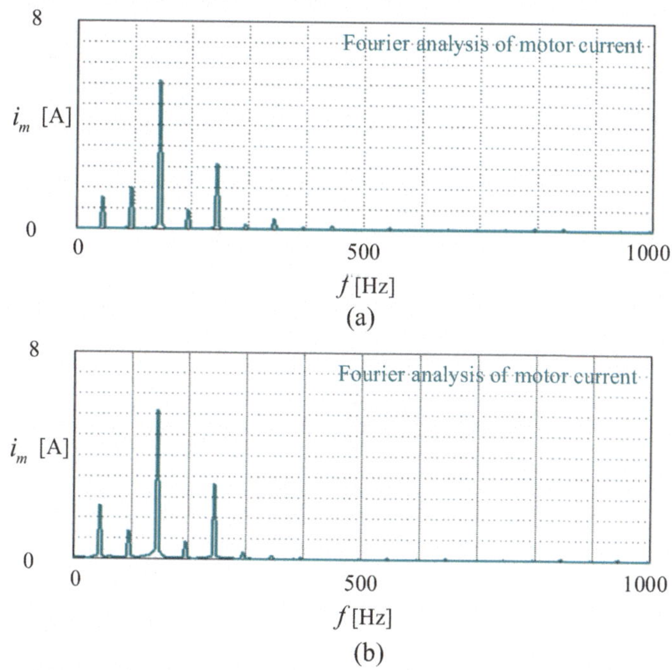

Fig. 3.10 Fourier analysis of the motor current. **a** Without the proposed method. **b** With the proposed method, $K_{dampP} = 0.15$

In order to validate the impact of the line impedance change on the active damping effect, experimental results are shown in Fig. 3.11 when the line impedance is 2.5 mH. Experimental waveforms from top to bottom are the grid voltage, the DC-link voltage, the grid current, and the motor current, respectively. The resonance between the line inductor and the DC-link capacitor is obvious. The resonant frequency is 710 Hz, which is higher than the one when the line impedance is 5mH. The DC-link voltage distortion is reduced, which can be seen from Fig. 3.11a. The determination of K_{dampP} can be realized by the method presented in Sect. 3.2.5, and the optimized value of K_{dampP} is 0.20. It can be seen from Fig. 3.11 that the proposed active damping method can effectively suppress the resonance as the line impedance changes.

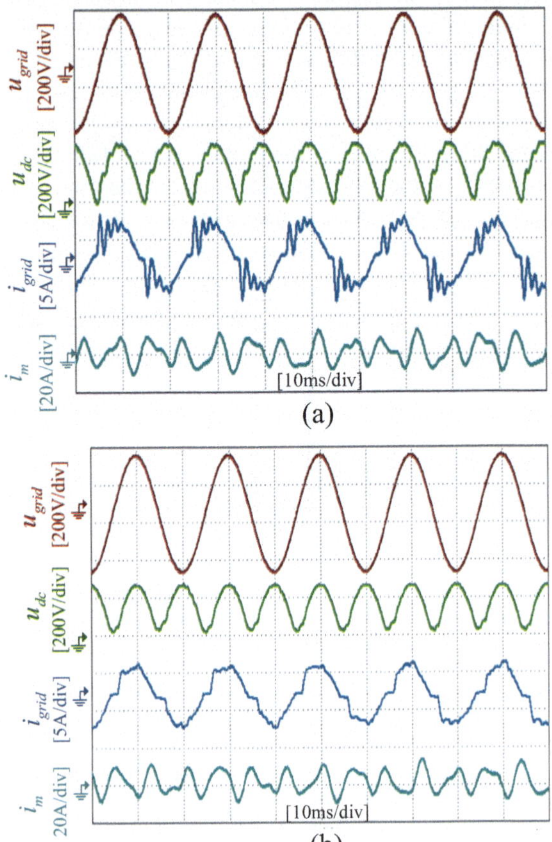

Fig. 3.11 Experimental waveforms of the grid voltage, DC-link voltage, grid current and motor current when the input inductance filter is 2.5 mH. **a** Without the proposed method. **b** With the proposed method, K_{dampP}= 0.2

3.3 Virtual Resistor Based Active Damping Control

3.3.1 *Different Configurations of Virtual Damping Resistor*

The virtual damping resistor could increase the system damping and suppress the LC resonance. Possible configurations of the virtual damping resistor are shown in Fig. 3.12. It can be seen that the damping resistor can be set in series with or in parallel to the line inductor and the film capacitor. Transfer functions under different locations of the damping resistor could be obtained and the Routh-Hurwitz criterion could be used to analyze the drive system stability. All the possible locations of the virtual damping resistor can change the coefficient of the characteristic polynomial and stabilize the drive system.

3.3.2 *Stability Analysis of Virtual Resistor Based Active Damping Control*

In order to evaluate the damping performance of the virtual damping resistor, Bode diagrams corresponding to the configurations in Fig. 3.12 are shown in Fig. 3.13 to analyze the drive system further. It can be seen from Fig. 3.13 that the LC resonance can be effectively suppressed by all the four possible solutions, and the resonant peak values are reduced by applying the virtual damping resistor. The damping performance is different with the four possible solutions. In Fig. 3.12a, the virtual damping resistor is parallel to the line inductor, and the impedance reduces in the high-frequency domain compared with the one without the virtual damping resistor, which decreases the attenuation of the magnitude when the frequency is higher than ω_{res} as shown in Fig. 3.13a. In Fig. 3.12b, the damping resistor is in series with the line inductor and the performance of the drive system in the high-frequency domain is the same regardless of the damping resistor, because the impedance of the damping resistor is relatively small compared with the line inductor when the frequency is higher than ω_{res}. The magnitude of Bode diagrams in Fig. 3.13b becomes higher when the frequency is lower than ω_{res}, which could improve the drive system performance. In Fig. 3.12c, the damping resistor is in parallel to the film capacitor, and the magnitude of Bode diagrams reduces when the frequency is lower than ω_{res} in Fig. 3.13c, which is not preferred to the ones in Fig. 3.13b. In Fig. 3.12d, the damping resistor is in series with the film capacitor, and the magnitude of Bode diagrams increases when the frequency is higher than ω_{res} in Fig. 3.13d, which decreases the drive system performance.

It can be concluded that the performance of the drive system in Fig. 3.13b is better than other three configurations, because the magnitude of Bode diagram owns the higher gain in the low-frequency domain and the higher attenuation in the high-frequency domain, which benefits the drive system performance. Generally, as for the damping resistor in series with the line inductor, the magnitude in the high-frequency

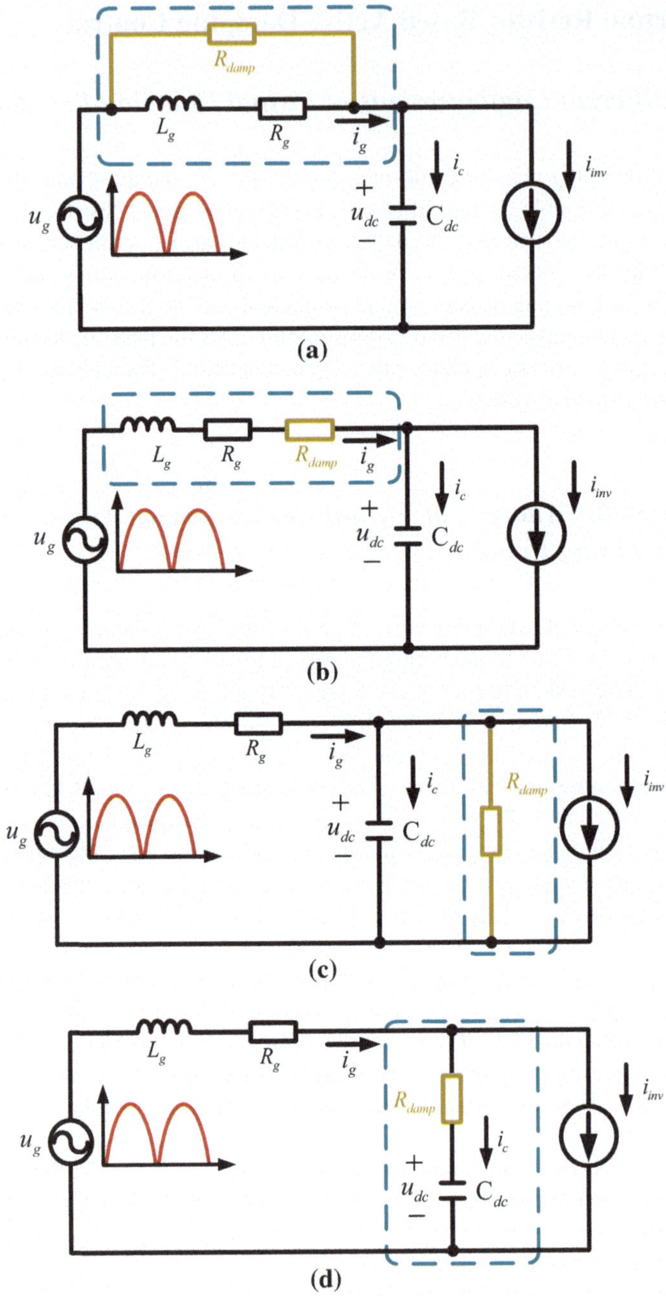

Fig. 3.12 Configurations of the virtual damping resistor of the drive system. **a** In parallel to the line inductor. **b** In series with the line inductor. **c** In parallel to the film capacitor. **d** In series with the film capacitor

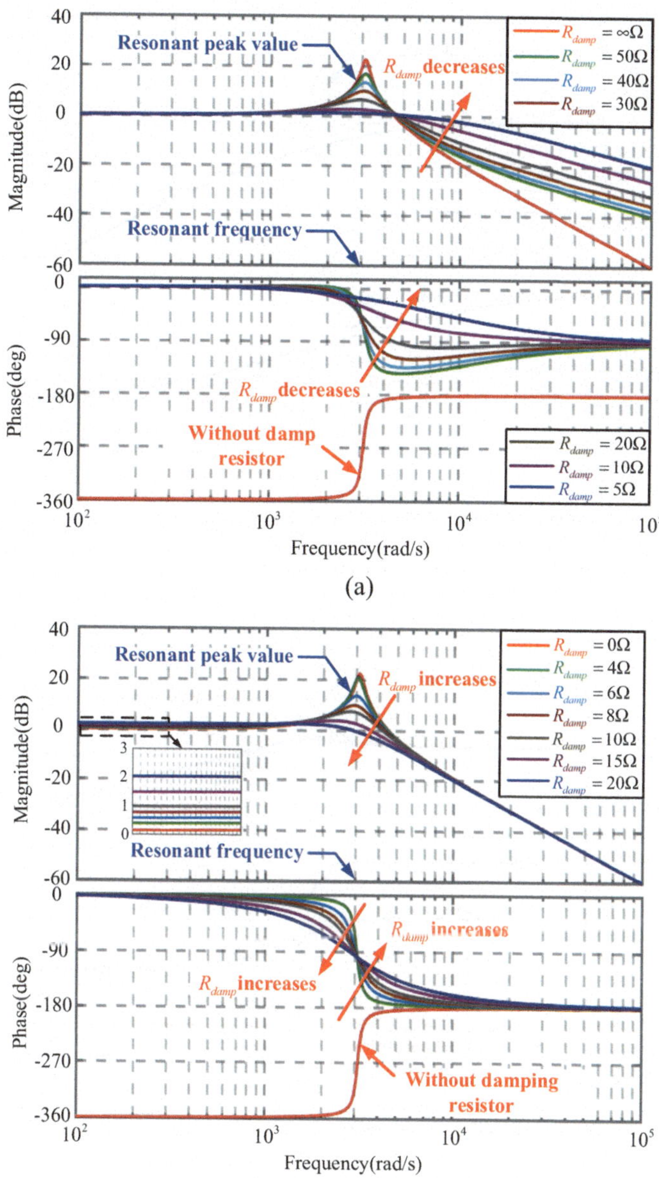

Fig. 3.13 Bode diagram of the drive system with different configurations of the virtual damping resistor. **a** In parallel to the line inductor. **b** In series with the line inductor. **c** In parallel to the film capacitor. **d** In series with the film capacitor

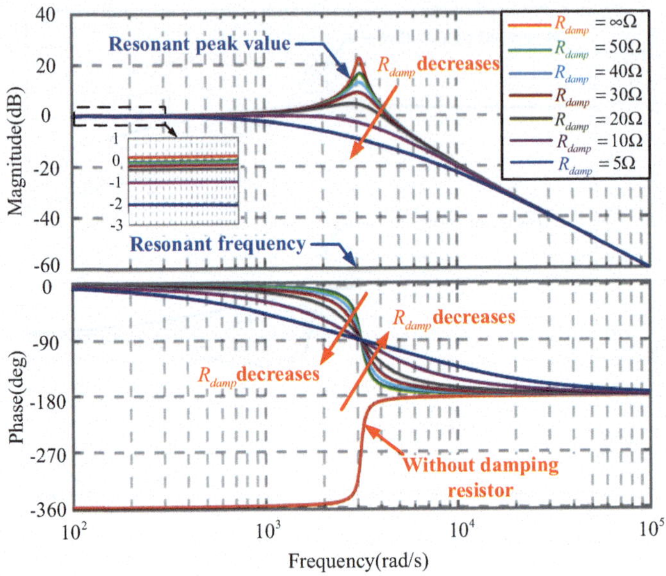

(c)

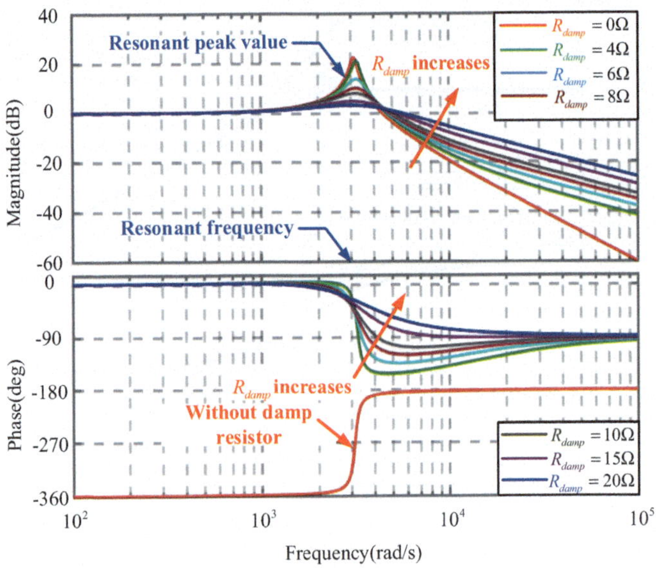

(d)

Fig. 3.13 (continued)

domain will increase when the load owns the positive input-impedance. However, the magnitude in the high-frequency domain decreases because the load is the specified CPL, which owns the negative input impedance. Hence the virtual damping resistor in series with the line inductor is applied to suppress the LC resonance and stabilize the drive system.

3.4 Inductor Current Feedback Based Active Damping Control Method

3.4.1 Realization of Inductor Current Feedback Control

As analyzed above, the optimal solution is to add the virtual damping resistor in series with the line inductor. Hence the inductor current, i.e., the grid current i_g, is applied to establish the feedback loop in Fig. 3.14a, where K_{vrc} represents the coefficient of $\tilde{u}_{dc}$ in (3.4). The inductor current feedback (ICF) method could effectively emulate the damping performance the same as the virtual damping resistor.

As shown in Fig. 3.14a, the block diagram is difficult to be realized in actual application, because the summing point of the damping resistor in Fig. 3.14a cannot be realized by the control algorithm. Hence it is essential to make the ICF method to be achievable in the actual application. It can be seen from Fig. 3.14a that the possible feedback signals to replace the inductor current i_g include the capacitor current i_c, the DC-link voltage u_{dc}, and the inverter current i_{inv}. The solution is to replace the ICF loop by the DC-link voltage feedback loop, because the DC-link voltage sensor is a general part in the conventional drive system for the voltage protection, and it does not need additional sensors or operations compared with i_{inv} and i_c.

The equivalent block diagram of Fig. 3.14a is shown in Fig. 3.14b, and the DC-link voltage feedback loop is established in Fig. 3.14c. It can be seen that the summing point of the DC-link voltage feedback loop is between the grid voltage and the line inductor, which is also difficult to be realized in actual application. Hence the summing point is moved backward as shown in Fig. 3.14d, and the feedback loop can be regarded as adding a specific damping current i_{damp} to the branch of the inverter. The realization of i_{damp} could be emulated by the motor voltage command and it will be introduced in the last part in this section.

3.4.2 Compensation of Distorted Grid Voltage

In actual applications, the grid voltage usually contains the background harmonics of the power grid, and the grid current is closely related to the grid voltage in the reduced DC-link capacitance motor drive system. Hence it is essential to compensate the effect of the distorted grid voltage on the grid current.

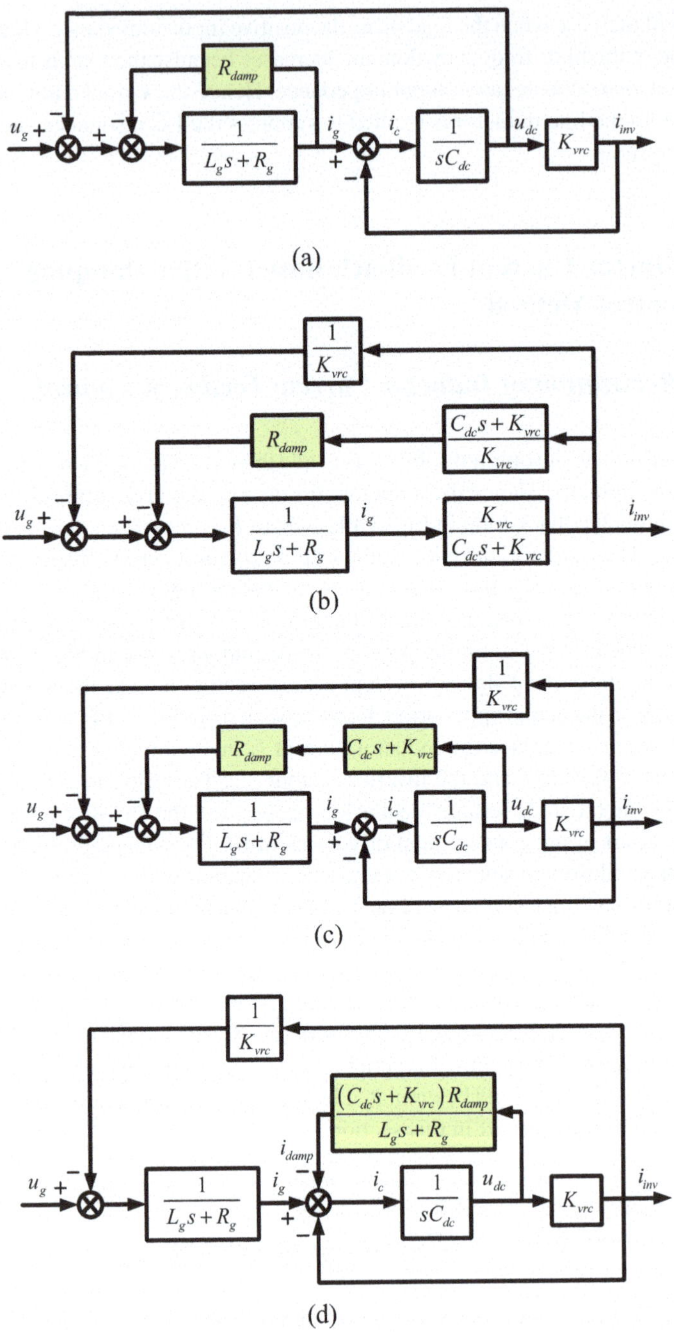

Fig. 3.14 Equivalent transformation of the ICF loop

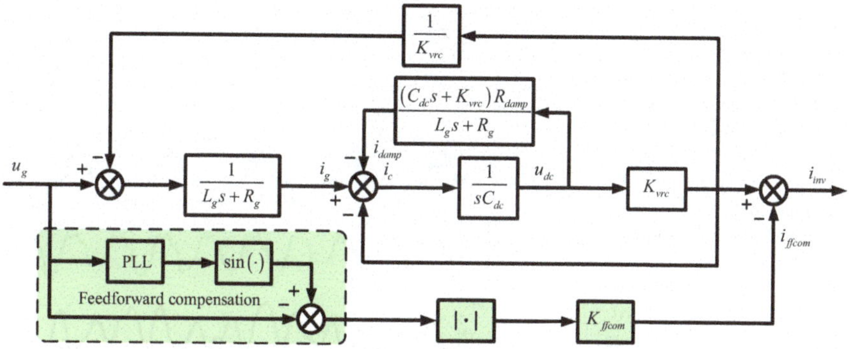

Fig. 3.15 Feedforward compensation of the grid voltage distortion

The feedforward compensation method is applied on the basis of the ICF-based active damping method, which is shown in Fig. 3.15. It can be seen that harmonics of the grid voltage are detected by the difference of the ideal grid voltage, which is generated by the PLL and the actual grid voltage. The grid voltage is regarded as a particular voltage source with the absolute value. Hence the absolute operation is applied in the feedforward compensation method. The coefficient K_{ffcom} of the feedforward compensation current i_{ffcom} is presented as

$$K_{ffcom} = \frac{K_{vrc}}{L_g C_{dc} s^2 + \left(R_g C_{dc} + R_{damp} C_{dc} + K_c L_g\right)s + K_c\left(R_{damp} + R_g\right) + 1}. \quad (3.27)$$

The feedforward compensation current i_{ffcom} is directly added to i_{inv}, which will be generated by the inverter based on the normal motor operation. The application of the feedforward compensation method can overcome the effect of the harmonics of the grid voltage on the drive system, and the performance of the grid current could be improved further.

3.4.3 Experimental Results

Figure 3.16 shows the experimental results of the reduced DC-link capacitance motor drive system without and with the proposed ICF-based active damping method and the feedforward compensation method. Waveforms from up to bottom are the grid voltage, the grid current, the DC-link voltage and the motor current, respectively. The Fourier analysis of the grid current is also shown to demonstrate the performance comparison. Without the proposed control method, it can be seen from Fig. 3.16a that harmonics of the DC-link voltage and the grid current are obvious, which are caused by the LC resonance between the line input inductor and the film capacitor. The

Fig. 3.16 Experimental waveforms of the grid voltage, the grid current, the DC-link voltage, and the motor current when the motor speed is 3000 rpm. **a** Without the proposed method. **b** With the ICF-based active damping method. **c** With the ICF-based active damping method and the feedforward compensation of grid voltage distortion

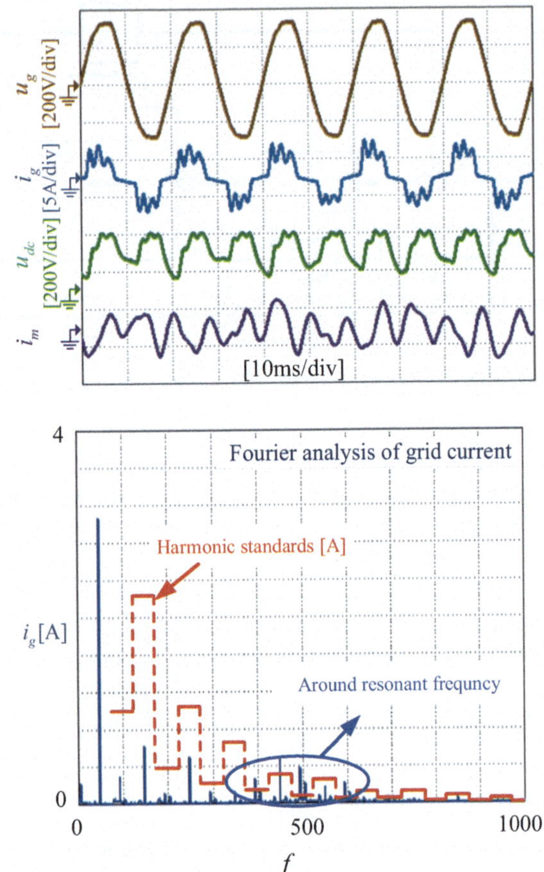

(a)

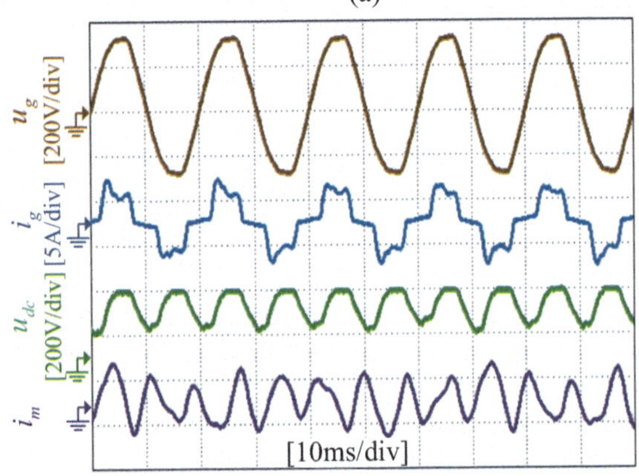

Fig. 3.16 (continued)

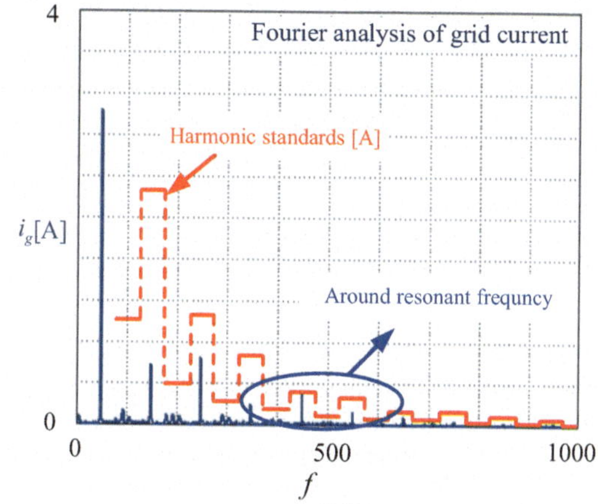

(b)

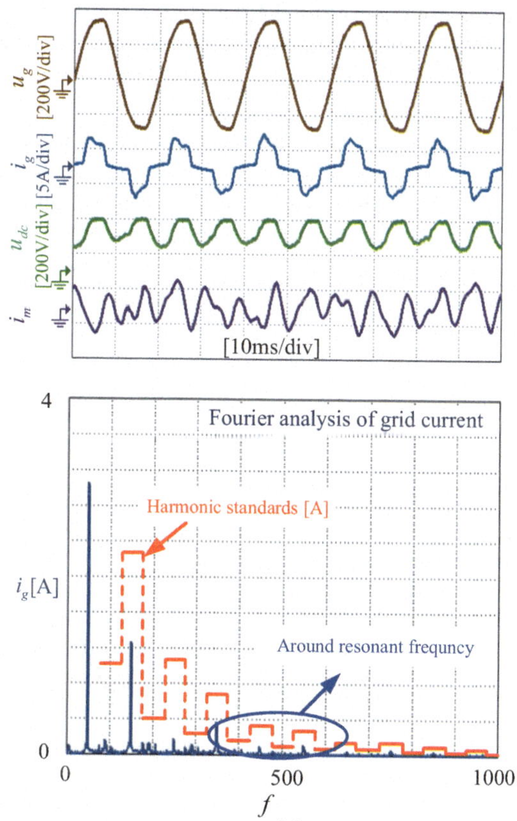

(c)

Fourier analysis of the grid current cannot meet the requirements of the EN61000-3-2. Harmonics around the resonant frequency are severe due to the peak value as shown in Fig. 3.13, and amplitudes of harmonics at 450 Hz, 500 Hz, and 550 Hz are 0.5A, 0.4A, and 0.18A, respectively. Hence it is essential to be suppressed considering its pollution to the power grid. With the ICF-based active damping method, it can be seen from Fig. 3.16b that harmonics of the DC-link voltage and the grid current are reduced due to the suppression of the LC resonance. The Fourier analysis of the grid current at 450 Hz, 500 Hz, and 550 Hz are 0.35A, 0.02A, and 0.12A, respectively, which are far lower than that in Fig. 3.16a. It can be seen from Fig. 3.16a and b that the grid voltage contains the background harmonics of the power grid, which could cause additional harmonics of the grid current. The feedforward compensation method is applied in Fig. 3.16c. Harmonics of the grid current at 450 Hz, 500 Hz, and 550 Hz are 0.08A, 0.01A, and 0.09A, respectively, which are reduced further compared with the ones in Fig. 3.16b. Hence the negative effect of the distorted grid voltage can be compensated, and the Fourier analysis of the grid current can meet the requirements of the EN61000-3-2.

The feedforward compensation method is realized by the error of the ideal grid voltage generated by the PLL and the actual grid voltage. Meanwhile, the sum of the active damping current i_{damp} and the feedforward current i_{ff} is also shown in Fig. 3.17, which is emulated by the motor voltage. Waveforms of the motor voltage and current in the stationary frame are shown in Fig. 3.18 to show the performance of the drive system (the motor operates at 3000 rpm). Compared with the conventional motor drive system, the voltage and current are not ideal sinusoidal, which are affected by the fluctuated DC-link voltage, the motor load, and the additional voltage command generated by the proposed control method.

In order to verify the effectiveness of the proposed method further, the experimental results are shown in Fig. 3.19 when the motor speed is 4000 rpm. Due to the instability caused by the LC resonance, the drive system is unable to operate normally

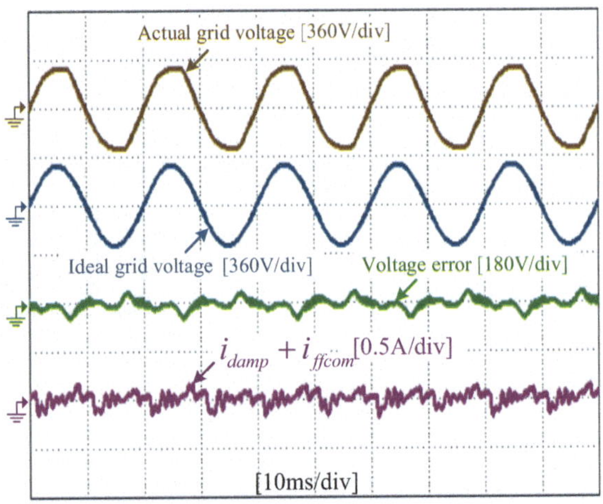

Fig. 3.17 Experimental waveforms of the actual grid voltage, the ideal grid voltage, the voltage error, and i_{damp}+ i_{ffcom} when the motor operates at 3000 rpm

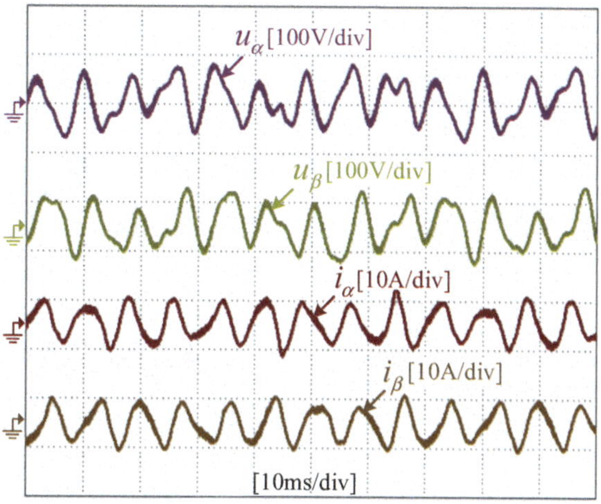

Fig. 3.18 Experimental waveforms of the voltage and current in the stationary frame when the motor operates at 3000 rpm

when the motor speed is higher than 4000 rpm, and harmonics of the grid current extremely exceed the requirements of EN61000-3-2. It can be seen from Fig. 3.19 that harmonics of the grid current are reduced with the application of the proposed ICF-based active damping method and the feedforward compensation of the distorted grid voltage. The performance of the proposed method at different operation points is shown in Fig. 3.20. It can be seen that magnitudes of harmonics around the resonant frequency are reduced obviously by applying the ICF-based active damping method and the feedforward compensation method, which benefit the application of the reduced DC-link capacitance motor drive system.

3.5 Summary

The active damping control method for the reduced DC-link capacitance IPMSM drive system is proposed in this chapter. The LC resonance between the line inductor and the DC-link capacitor is obvious due to the reduced DC-link capacitance. Meanwhile, the negative input-impedance weakens the drive system stability. The proposed DC-link voltage feedback based active damping control method can suppress the resonance and improve the stability of the drive system. Meanwhile, all possible configurations of the damping resistor are analyzed and the optimal solution is equivalent to the virtual damping resistor in series with the line inductor. A feedforward compensation scheme is proposed to attenuate the negative effect of the distorted grid voltage, which could reduce harmonics of the grid current further. The application of the proposed control method could reduce harmonics of the grid current to meet the standard requirements and benefit the application of the reduced DC-link capacitance motor drive system.

Fig. 3.19 Experimental waveforms of the grid voltage, the grid current, the DC-link voltage, and the motor current when the motor speed is 4000 rpm. **a** Without the proposed method. **b** With the ICF-based active damping method. **c** With the ICF-based active damping method and the feedforward compensation of the grid voltage distortion

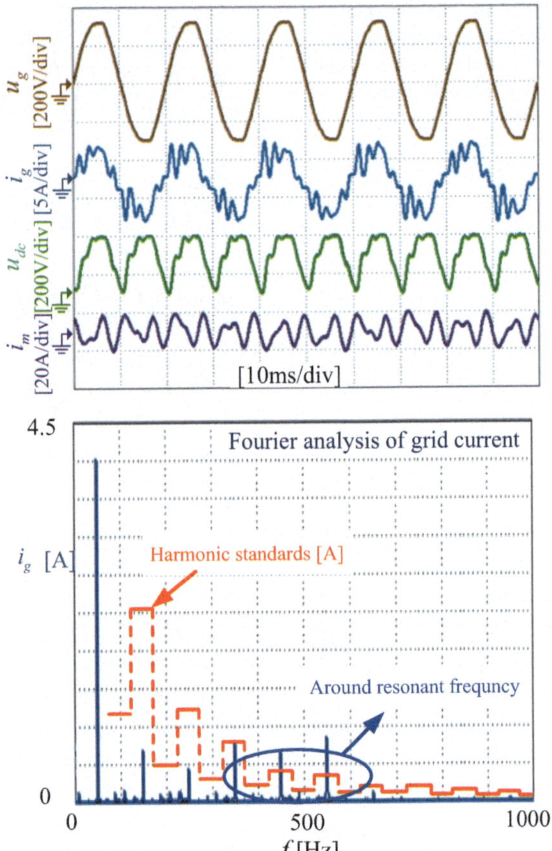

(a)

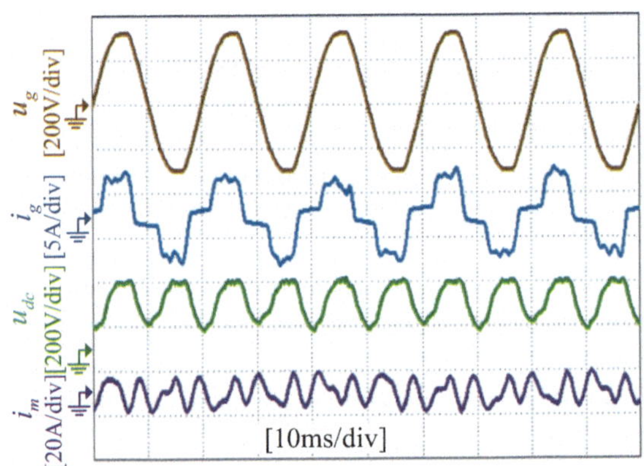

Fig. 3.19 (continued)

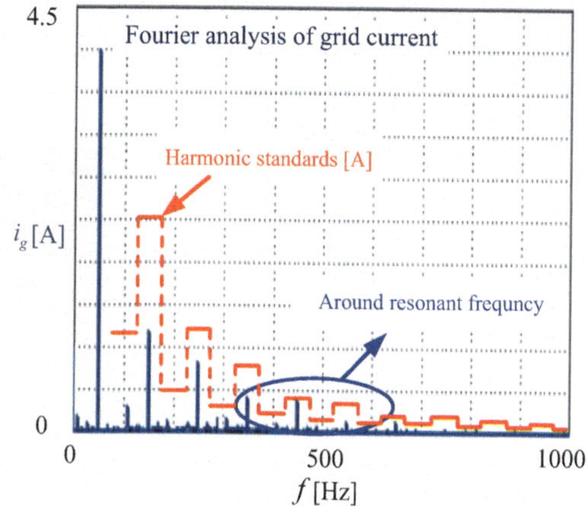

(b)

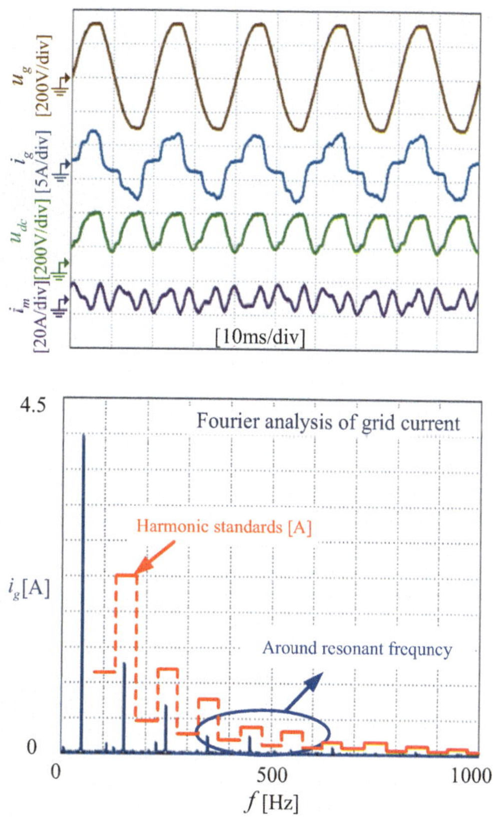

(c)

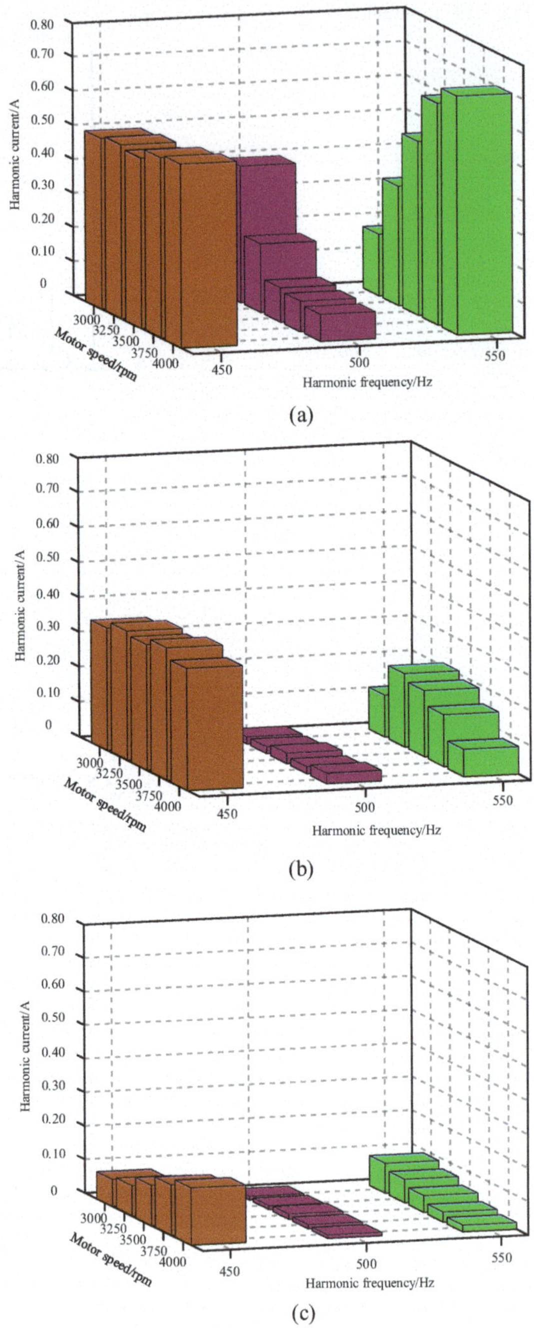

Fig. 3.20 Experimental results of grid current harmonics at different operating points. **a** Without the proposed method. **b** With the ICF-based active damping method. **c** With the ICF-based active damping method and the feedforward compensation of the grid voltage distortion

References

1. D. Wang, K. Lu, P.O. Rasmussen, L. Mathe, Y. Feng, F. Blaabjerg, Voltage modulation using virtual positive impedance concept for active damping of small DC-link drive system. IEEE Trans. Power Electron. **33**(12), 10611–10621 (2018)
2. L. Mathe, L. Török, D. Wang, D. Sera, Resonance reduction for AC drives with small capacitance in the DC link. IEEE Trans. Ind. Appl **53**(4),3814–3820 (2017)
3. R. Maheshwari, S. Munk-Nielsen, K. Lu, An active damping technique for small DC-link capacitor based drive system. IEEE Trans. Ind. Inform. **9**(2), 848–858 (2013)
4. L. Mathe, H.R. Andersen, R. Lazar, M. Ciobotaru, DC-link compensation method for slim DC-link drives fed by soft grid, in *IEEE International Symposium on Industrial Electronics* (2010), pp. 1236–1241
5. W. Lee, S. Sul, DC-link voltage stabilization for reduced DC-link capacitor inverter. IEEE Trans. Ind. Appl. **50**(1), 404–414 (2014)
6. A.B. Awan, S. Pierfederici, B. Nahid-Mobarakeh, F. Meibody-Tabar, Active stabilization of a poorly damped input filter supplying a constant power load, in *IEEE Energy Conversion Congress and Exposition* (2009), pp. 2991–2997
7. S.D. Sudhoff, K.A. Corzine, S.F. Glover, H.J. Hegner, H.N. Robey, DC link stabilized field oriented control of electric propulsion systems. IEEE Trans. Energy Convers. **13**(1), 27–33 (1998)
8. N. Zhao, G. Wang, D. Xu, D. Xiao, An active damping control method for reduced DC-link capacitance IPMSM drives. IEEE Trans. Ind. Electron. **65**(3), 2057–2068 (2018)
9. N. Zhao, G. Wang, R. Zhang, B. Li, Y. Bai, D. Xu, Inductor current feedback active damping method for reduced DC-link capacitance IPMSM drives. IEEE Trans. Power Electron. **34**(5), 4558–4568 (2019)
10. K. Pietilainen, L. Harnefors, A. Petersson, H.P. Nee, DC-Link Stabilization and Voltage Sag Ride-Through of Inverter Drives. IEEE Trans. Ind. Electron. **53**(4), 1261–1268 (2006)

Chapter 4
Impedance Model Based Stability Control

The instability of the drive system is caused by the interaction of the LC resonance and the constant power load (CPL), hence the active damping methods have been applied to damp the resonance and stabilize the drive system. In the previous chapter, the grid current harmonics suppression methods and the active damping methods could effectively improve the grid current performance and stabilize the drive system. However, the relationship between the system stability and the suppression of grid current harmonics needs to be explained comprehensively. Hence the impedance model of the reduced DC-link capacitance motor drive system is applied in this chapter to evaluate the relationship between the drive system stability and the grid current performance.

4.1 Impedance Modeling of PMSM

The reduced DC-link capacitance PMSM drives could be simplified as the block diagram shown in Fig. 4.1. The small signal of the motor voltage is applied to evaluate the DC-link voltage fluctuation on the actual *d-q* axis voltages [1–3], which could be presented as

$$\begin{bmatrix} \Delta u_d \\ \Delta u_q \end{bmatrix} = \frac{\Delta u_{dc}}{u_{dc,0}} \begin{bmatrix} u_{d,0} \\ u_{q,0} \end{bmatrix} + \begin{bmatrix} \Delta u_{dref} \\ \Delta u_{qref} \end{bmatrix} \tag{4.1}$$

where Δu_d, Δu_q, Δu_{dref}, Δu_{qref}, Δu_{dc}, $u_{d,0}$, and $u_{q,0}$ are the small signals of the actual *d-q* axis motor voltages, the small signal of the *d-q* axis voltages reference generated by the current controllers, the small signal of the DC-link voltage, the average *d-q* axis voltages, respectively. It means that the small signals of the actual *d-q* axis voltages are equal to the small signals of the *d-q* axis voltages reference when the DC-link voltage is constant. Meanwhile, the effect of Δu_{dc} on Δu_d and Δu_q is influenced by $u_{d,0}/u_{dc,0}$ and $u_{q,0}/u_{dc,0}$, respectively.

© The Author(s), under exclusive license to Springer Nature Singapore Pte Ltd. 2021
G. Wang et al., *Reduced DC-link Capacitance AC Motor Drives*,
https://doi.org/10.1007/978-981-15-8566-1_4

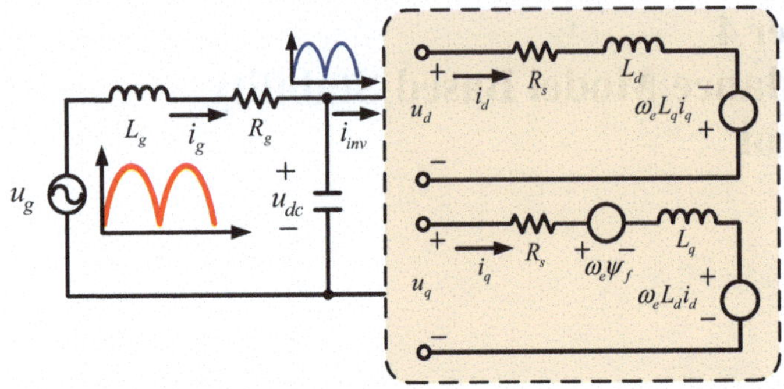

Fig. 4.1 Impedance model of the drive system

The motor voltages in the synchronous frame are presented as

$$\begin{bmatrix} u_d \\ u_q \end{bmatrix} = \begin{bmatrix} R_s + L_d s & -\omega_e L_q \\ \omega_e L_d & R_s + L_q s \end{bmatrix} \begin{bmatrix} i_d \\ i_q \end{bmatrix} + \begin{bmatrix} 0 \\ \omega_e \psi_f \end{bmatrix} \tag{4.2}$$

Hence the small signals of (4.2) are presented as

$$\begin{bmatrix} \Delta u_d \\ \Delta u_q \end{bmatrix} = \begin{bmatrix} R_s + L_d s & -\omega_e L_q \\ \omega_e L_d & R_s + L_q s \end{bmatrix} \begin{bmatrix} \Delta i_d \\ \Delta i_q \end{bmatrix} \tag{4.3}$$

where Δi_d and Δi_q are the small signals of the actual *d*-*q* axis currents, respectively. The relationship between the actual motor currents and the actual motor voltages is shown in (4.3).

The voltage references of *d*-*q* axis in the synchronous frame are generated by the current controllers, which are presented as

$$\begin{bmatrix} u_d^* \\ u_q^* \end{bmatrix} = \begin{bmatrix} -G_d & -\omega_e L_q \\ \omega_e L_d & -G_q \end{bmatrix} \begin{bmatrix} i_d \\ i_q \end{bmatrix} + \begin{bmatrix} G_d i_d^* \\ G_q i_q^* + \omega_e \psi_f \end{bmatrix} \tag{4.4}$$

where G_d and G_q are the current controllers of the *d*-*q* axis in the synchronous frame, respectively. The current controllers could be presented as

$$G_d = K_{idP} + \frac{K_{idI}}{s}, G_q = K_{iqP} + \frac{K_{iqI}}{s} \tag{4.5}$$

where K_{idP}, K_{iqP}, K_{idI}, and K_{iqI} are the proportional and integral gain of G_d, and G_q, respectively.

The small signal of (4.4) could be presented as

$$\begin{bmatrix} \Delta u_d^* \\ \Delta u_q^* \end{bmatrix} = \begin{bmatrix} -G_d & -\omega_e L_q \\ \omega_e L_d & -G_q \end{bmatrix} \begin{bmatrix} \Delta i_d \\ \Delta i_q \end{bmatrix} \tag{4.6}$$

Combing (4.1), (4.3), and (4.6), the relationship between Δi_d, Δi_q and Δu_{dc} could be presented as

$$\begin{bmatrix} \Delta i_d \\ \Delta i_q \end{bmatrix} = \frac{\Delta u_{dc}}{u_{dc,0}} \begin{bmatrix} \frac{u_{d,0}}{R_s + L_d s + G_d} \\ \frac{u_{q,0}}{R_s + L_q s + G_q} \end{bmatrix}. \tag{4.7}$$

Ignoring the power loss of the inverter, the inverter power is equal to the motor power, which could be presented as

$$u_{dc} i_{inv} = \frac{3}{2}\left(u_d i_d + u_q i_q\right). \tag{4.8}$$

The small signal of (4.8) is presented as

$$\Delta u_{dc} i_{inv,0} + u_{dc,0} \Delta i_{inv} = \frac{3}{2}\left(\Delta u_d i_{d,0} + u_{d,0} \Delta i_d + \Delta u_q i_{q,0} + u_{q,0} \Delta i_q\right) \tag{4.9}$$

where $i_{dc,0}$, $i_{d,0}$, and $i_{q,0}$ are the average value of i_{dc}, i_d and i_q, respectively. Hence the input admittance $Y_m(s)$ of PMSM in the reduced DC-link capacitance could be presented as

$$\begin{aligned} Y_m(s) = & -\frac{i_{inv,0}}{u_{dc,0}} + \frac{3}{2}\frac{u_{d,0}}{u_{dc,0}^2}\frac{u_{d,0} + \omega_e L_d i_{q,0} + L_d i_{d,0} s + i_{d,0} R_s}{R_s + L_d s + G_d} \\ & + \frac{3}{2}\frac{u_{q,0}}{u_{dc,0}^2}\frac{u_{q,0} - \omega_e L_q i_{d,0} + L_q i_{q,0} s + i_{q,0} R_s}{R_s + L_q s + G_q}. \end{aligned} \tag{4.10}$$

Bode diagrams of $Y_m(s)$ with different bandwidths of the current controller (200 Hz and 800 Hz) are shown in Fig. 4.2. It can be seen that the motor input admittance decreases rapidly when the current controller is with lower bandwidth. It means the drive system is less sensitive to the disturbance signals with a lower bandwidth current controller, which could enhance the drive system stability. The proposed motor input admittance model is consistent with the motor input admittance models applied in [4–6].

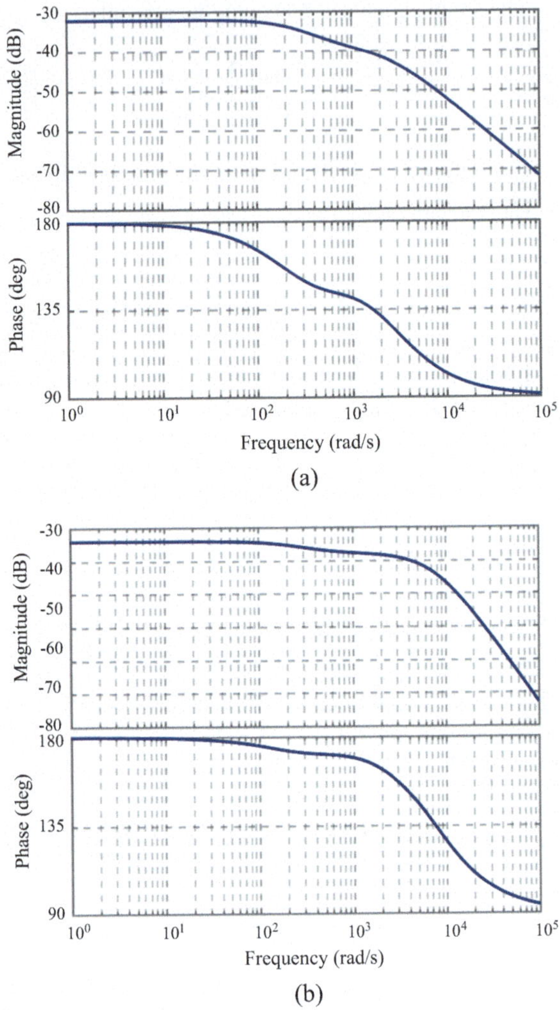

Fig. 4.2 Bode diagram of the motor input admittance Ym(s) with different bandwidth of current controllers. **a** The bandwidth of the current controller is 200 Hz. **b** The bandwidth of the current controller is 800 Hz

4.2 System Performance Evaluation

4.2.1 System Stability Analysis

The drive system could be regarded as a cascade system. The first stage is the LC input filter and the second stage is the PMSM drive. The stability of the cascade system could be evaluated by the ratio of the input admittance of the PMSM and the output admittance of the LC filter ($Y_{LC}(s)/Y_m(s)$). Meanwhile, the grid current harmonics caused by the LC resonance could also be analyzed by the input impedance from the grid side.

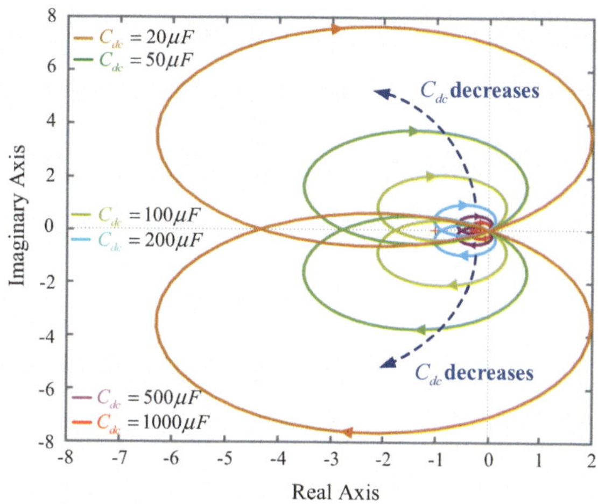

Fig. 4.3 Nyquist plots of $Y_{LC}(s)/Y_m(s)$ with different values of C_{dc}

The output admittance of the LC filter $Y_{LC}(s)$ could be presented as

$$Y_{LC}(s) = \frac{L_g C_{dc} s^2 + R_g C_{dc} s + 1}{L_g s + R_g}. \tag{4.11}$$

Nyquist plots of $Y_{LC}(s)/Y_m(s)$ with different values of the DC-link capacitance C_{dc} are shown in Fig. 4.3. It can be seen that the drive system stability reduces with the decrease of C_{dc}. Hence the stability issue is essential to be concerned in the reduced DC-link capacitance motor drive system, because of the remarkably reduced DC-link capacitance of the applied film capacitor.

4.2.2 *Analysis of Grid Current Harmonics*

The input impedance from the grid side $Z_g(s)$ is applied to evaluate the grid current performance of the drive system, which can be presented as

$$Z_g(s) = L_g s + R_g + \frac{1}{sC_{dc} + Y_m(s)}. \tag{4.12}$$

Bode diagrams of $Z_g(s)$ with different values of C_{dc} are shown in Fig. 4.4. It can be seen that the resonant frequency increases as C_{dc} decreases and the impedance changes sharply around the resonant frequency, which could stimulate remarkable harmonics of the grid current. Hence it is essential to compensate the sharply change of the impedance value to improve the grid current performance.

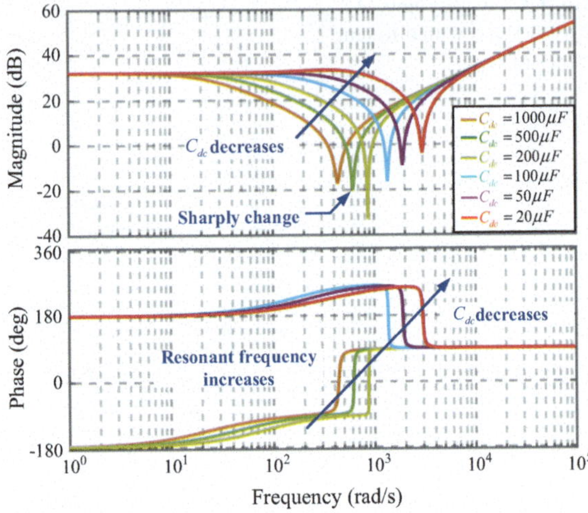

Fig. 4.4 Bode diagrams of $Z_g(s)$ with different values of C_{dc}

4.3 DC-Link Voltage Feedback Stability Control Method

4.3.1 *DC-Link Voltage Feedback Based Stability Control Method*

The conventional stabilization control method based on the DC-link voltage feedback is widely applied in the cascade drive system [4–6]. The DC-link voltage is processed by a band-pass filter $M_{dcv}(s)$ and applied to the motor voltage. The band-pass filter could be presented as

$$M_{dcv}(s) = \frac{2\xi_{dcv}\omega_{dcvc}s}{s^2 + 2\xi_{dcv}\omega_{dcvc}s + \omega_{dcvc}^2} \tag{4.13}$$

where ξ_{dcv} and ω_{dcvc} are the damping ratio and the operating frequency, respectively. ω_{dcvc} is equal to the resonant frequency of the LC filter ω_{res}.

After applying the conventional DC-link voltage feedback based stabilization control method, the q-axis voltage reference could be presented as

$$u_q^* = G_q\left(i_q^* - i_q\right) + K_{vv}M_{dcv}(s)u_{dc} + \omega_e L_d i_d + \omega_e \psi_f \tag{4.14}$$

where K_{vv} is the gain of the conventional stabilization control method.

4.3.2 *System Stability Analysis*

According to the previous analysis, Δi_q could be derived as

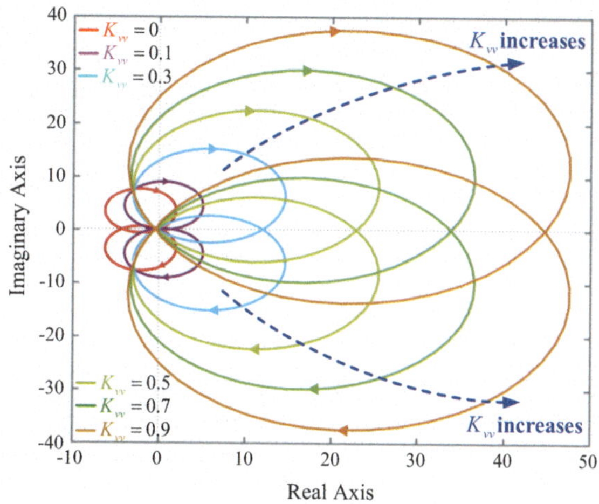

Fig. 4.5 Nyquist plots of $Y_{LC}(s)/Y_{mv}(s)$ with different values of K_{vv}

$$\Delta i_q = \frac{\Delta u_{dc}}{u_{dc,0}} \frac{u_{q,0}}{R_s + L_q s + G_q} + \frac{K_{vv} M_{dcv}(s) \Delta u_{dc}}{R_s + L_q s + G_q}. \tag{4.15}$$

Hence the input admittance of the PMSM $Y_{mv}(s)$ after applying the conventional stabilization control method could be presented as

$$\begin{aligned} Y_{mv}(s) &= -\frac{i_{dc,0}}{u_{dc,0}} + \frac{3}{2}\frac{u_{d,0}}{u_{dc,0}^2} \frac{u_{d,0} + \omega_e L_d i_{q,0} + L_d i_{d,0} s + i_{d,0} R_s}{R_s + L_d s + G_d} \\ &+ \frac{3}{2}\frac{u_{q,0}}{u_{dc,0}^2} \frac{u_{q,0} - \omega_e L_q i_{d,0} + L_q i_{q,0} s + i_{q,0} R_s}{R_s + L_q s + G_q} \\ &+ \frac{3}{2}\frac{K_{vv} M_{dcv}(s)}{u_{dc,0}} \frac{u_{q,0} - \omega_e L_q i_{d,0} + L_q i_{q,0} s + i_{q,0} R_s}{R_s + L_q s + G_q} \\ &= Y_m(s) + Y_{vv}(s) \end{aligned} \tag{4.16}$$

where $Y_{vv}(s)$ is the additional admittance caused by the conventional stabilization control method.

Nyquist plots of $Y_{LC}(s)/Y_{mv}(s)$ with different values of K_{vv} are shown in Fig. 4.5. It can be seen that the drive system stability is enhanced by the stabilization control method, and the system becomes stable when K_{vv} is higher than 0.3.

4.3.3 *Analysis of Grid Current Harmonics*

The grid current performance is another important issue in the reduced DC-link capacitance motor drives, and the effect of the stabilization control method on the grid current is essential to be analyzed. The input impedance from the grid side after

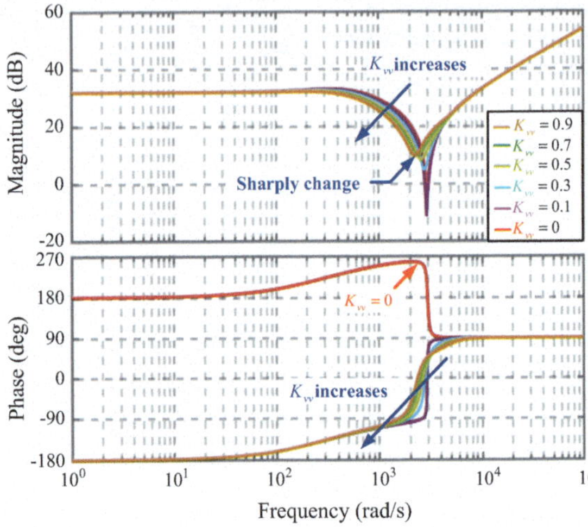

Fig. 4.6 Bode diagrams of $Z_{gv}(s)$ with different values of K_{vv}

applying the conventional stabilization control method $Z_{gv}(s)$ could be presented as

$$Z_{gv}(s) = L_g s + R_g + \frac{1}{sC_{dc} + Y_{mv}(s)}. \tag{4.17}$$

In order to evaluate the grid current performance, Bode diagrams of $Z_{gv}(s)$ are shown in Fig. 4.6. It can be seen that the performance of the LC resonance suppression is limited by applying the conventional stabilization control method. The sharply change of the impedance value caused by the LC resonance also exists, which is not the optimal suppression performance of the grid current harmonics.

4.4 Grid Current Feedback Based Stabilization Control Method

4.4.1 Principle of the Grid Current Feedback Based Stabilization Control Method

As analyzed in the previous section, the stability issue of the reduced DC-link capacitance motor drive system needs to be concerned, which is caused by the interaction between the LC resonance and the CPL. Meanwhile, the grid current harmonics caused by the LC resonance are also important concerns in actual applications. In this section, according to the impedance model, a grid current feedback based stabilization control method is introduced to stabilize the drive system and improve the grid current performance [7], which is shown in Fig. 4.7. The grid current is processed by a band-pass filter and applied to the q-axis stator voltage. The proportional gain

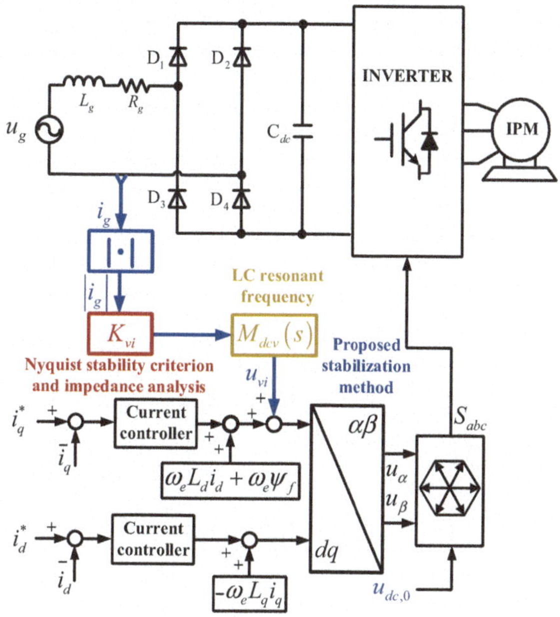

Fig. 4.7 Stabilization control method of reduced DC-link capacitance motor drives

K_{vi} is determined by the combination of the drive system stability evaluation and the grid current performance improvement, which are realized by the Nyquist stability criterion and the grid input impedance analysis, respectively.

4.4.2 System Stability Analysis

As for the grid current feedback based stabilization control method, the grid current i_g in (3.1) is applied to establish the feedback loop, which could be presented as

$$i_g = C_{dc} u_{dc} s + i_c. \tag{4.18}$$

The q-axis voltage reference after applying the stabilization control method could be presented as

$$u_q^* = G_q(i_q^* - i_q) + K_{vi} M_{dcv}(s) i_g + \omega_e L_d i_d + \omega_e \psi_f \tag{4.19}$$

where K_{vi} is the gain of the proposed stabilization control method. Then Δi_q could be derived as

$$\Delta i_q = \frac{\Delta u_{dc}}{u_{dc,0}} \frac{u_{q,0}}{R_s + L_q s + G_q} + \frac{K_{vi} M_{dcv}(s) \Delta i_g}{R_s + L_q s + G_q}. \tag{4.20}$$

Hence the input admittance of the PMSM after applying the proposed stabilization control method $Y_{mi}(s)$ could be presented as

$$Y_{mi}(s)=\frac{Y_m+\frac{3}{2}\frac{K_{vi}M_{dcv}(s)C_{dc}s}{u_{dc,0}}\frac{u_{q,0}-\omega_e L_q i_{d,0}+L_q i_{q,0}s+i_{q,0}R_s}{R_s+L_q s+G_q}}{1-\frac{3}{2}\frac{K_{vi}M_{dcv}(s)}{u_{dc,0}}\frac{u_{q,0}-\omega_e L_q i_{d,0}+L_q i_{q,0}s+i_{q,0}R_s}{R_s+L_q s+G_q}}. \tag{4.21}$$

Nyquist plots of $Y_{LC}(s)/Y_{mi}(s)$ with different values of K_{vi} are shown in Fig. 4.8. The overall Nyquist plots are shown in Fig. 4.8a and enlarged ones are shown in Fig. 4.8b. It can be seen that the stabilization control method could stabilize the drive system when K_{vi} is higher than 100.

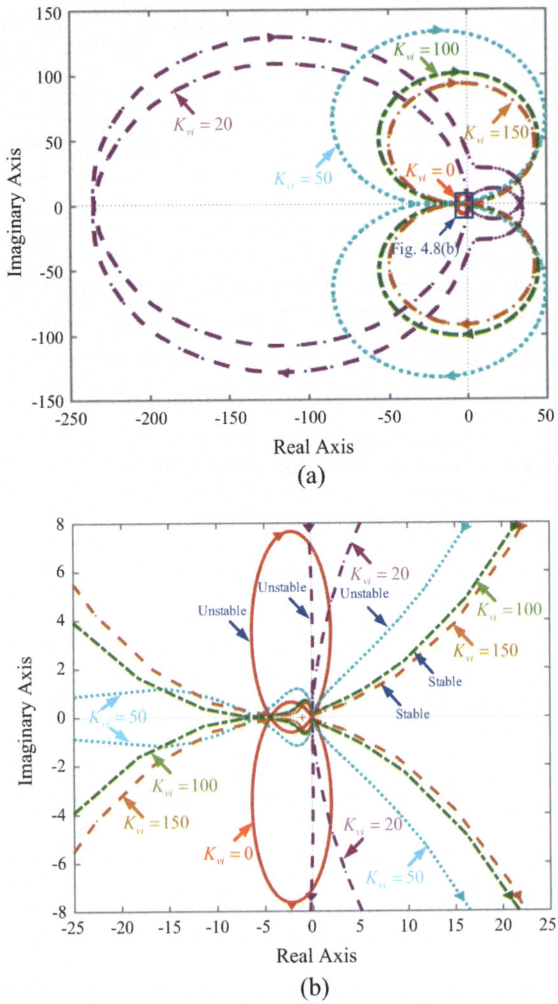

Fig. 4.8 Nyquist plots of $Y_{LC}(s)/Y_{mi}(s)$ with different values of K_{vi}. **a** Overall Nyquist plots. **b** Detailed ones

It can be seen that the drive system without the stabilization control method is unstable, and both the DC-link voltage feedback based stabilization control method and the grid current feedback based stabilization control method could stabilize the drive system by adjusting the values of K_{vi} and K_{vv}. As for the DC-link voltage feedback based stabilization method, the drive system becomes stable when K_{vv} is larger than 0.3. Meanwhile, the drive system also becomes stable with the grid current feedback based stabilization control method when K_{vi} is larger than 80. The range of K_{vi} is larger than K_{vv}, which means the drive system is less sensitive to the variety of K_{vi}.

4.4.3 Analysis of Grid Current Harmonics

As for the grid current feedback based stabilization control method, the input impedance from the grid side after applying the stabilization control method $Z_{gi}(s)$ could be presented as

$$Z_{gi}(s) = L_g s + R_g + \frac{1}{sC_{dc} + Y_{mi}(s)}. \tag{4.22}$$

Bode diagrams of $Z_{gi}(s)$ are shown in Fig. 4.9. Different from the conventional stabilization control method, the sharply change of the impedance value could be effectively suppressed by the proposed stabilization control method and the grid current performance could also be improved. Hence the proposed stabilization control method could stabilize the drive system and improve the grid current performance

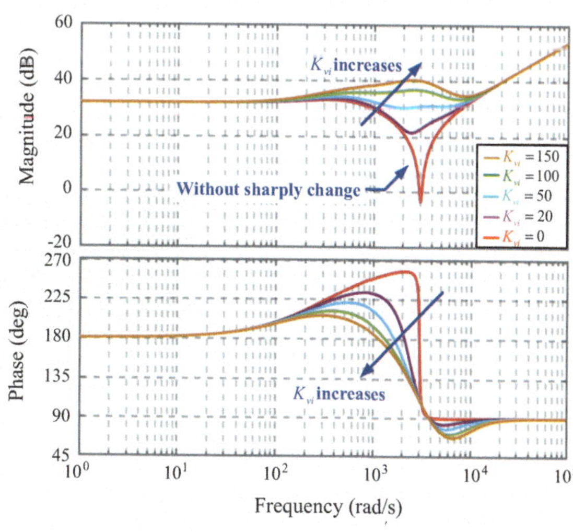

Fig. 4.9 Bode diagrams of $Z_{gi}(s)$ with different values of K_{vi}

simultaneously, which is the preferred stabilization control method for the reduced DC-link capacitance motor drives.

4.4.4 Experimental Results

The DC-link voltage u_{dc} and the grid voltage u_g before and after applying the proposed method are shown in Fig. 4.10. The PMSM operates at 3000 rpm. It can be seen that the peak value and the valley value of the DC-link voltage before applying the proposed stabilization control method are 325 V and 85 V, respectively. After applying the proposed stabilization control method, the peak value and the valley value change to 311 V and 100 V, respectively. Hence the DC-link voltage fluctuation is suppressed and the drive system stability is enhanced by applying the proposed stabilization control method.

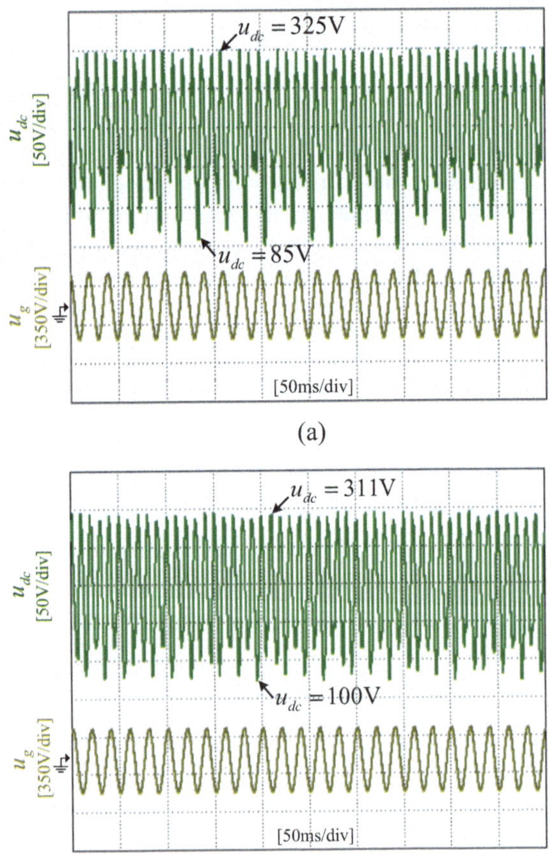

(b)

Fig. 4.10 Experimental waveforms of the DC-link voltage and the grid voltage when the compressor operates at 3000 rpm. **a** Without the proposed stabilization control method. **b** With the proposed stabilization control method

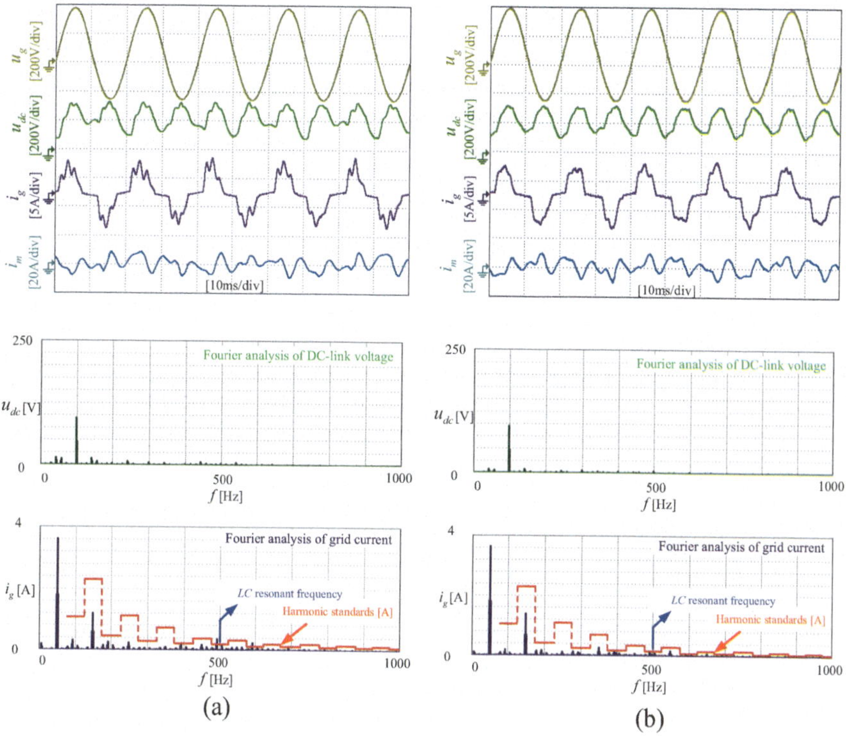

Fig. 4.11 Experimental waveforms of the grid voltage, DC-link voltage, grid current and motor current when the compressor operates at 3000 rpm. **a** Without the proposed stabilization control method. **b** With the proposed stabilization control method

Figure 4.11 shows the drive system performance before and after applying the proposed stabilization control method, corresponding to Fig. 4.11a and b, respectively. The compressor also operates at 3000 rpm. Waveforms from top to bottom are the grid voltage, the DC-link voltage, the grid current, and the motor current, respectively. Meanwhile, the Fourier analysis is carried out to analyze harmonics of the DC-link voltage and the grid current. It can be seen from Fig. 4.11a that the resonant components of the DC-link voltage and the grid current are obvious, which could weaken the drive system stability and stimulate the additional harmonics to the power grid. As shown in the Fourier analysis of the grid current, harmonics around the resonant frequency (500 Hz, $L_g = 5\text{mH}$ and $C_{dc} = 20\mu\text{F}$) could not meet the requirement of the IEC-61000-3-2, which limits the application of the reduced DC-link capacitance motor drive system. In order to stabilize the drive system and improve the grid current performance, the proposed stabilization control method is applied in Fig. 4.11b. It can be seen that the resonant component of the DC-link voltage is suppressed compared with the one in Fig. 4.11a. Meanwhile, harmonics of the grid current are suppressed remarkably as shown in the Fourier analysis.

In order to verify the effectiveness of the proposed stabilization control method, the compressor speed is set as 4000 rpm, and the experimental results are shown in Fig. 4.12. Waveforms from top to bottom are the grid voltage, the DC-link voltage, the grid current, and the motor current, respectively. It can be seen that the DC-link voltage could also be stabilized and harmonics of the grid current could be suppressed. Hence the proposed method could stabilize the drive system and suppress the grid current harmonics in a wide speed range.

Figure 4.13 shows the experimental results without and with the proposed stabilization control method. Waveforms from top to bottom are the q-axis voltage, the grid current, the absolute value of the grid current, and the voltage command generated by the proposed stabilization control method as shown in Fig. 4.7, respectively. It can be seen that the voltage command contains the resonant components, which are injected to the motor drive system. According to the classical control theory, the proper feedback signal could stabilize the system, and it is consistent with the

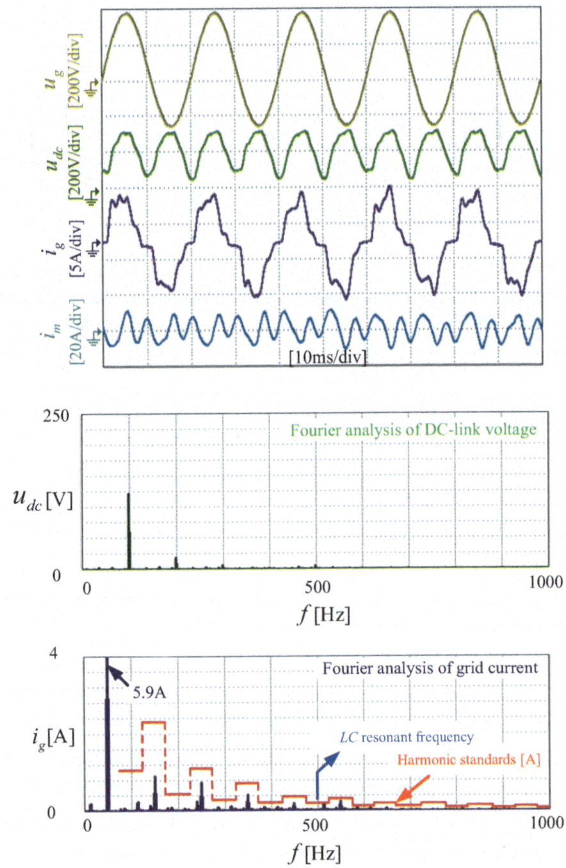

Fig. 4.12 Experimental waveforms of the grid voltage, DC-link voltage, grid current and motor current when the compressor operates at 4000 rpm (With the proposed stabilization control method)

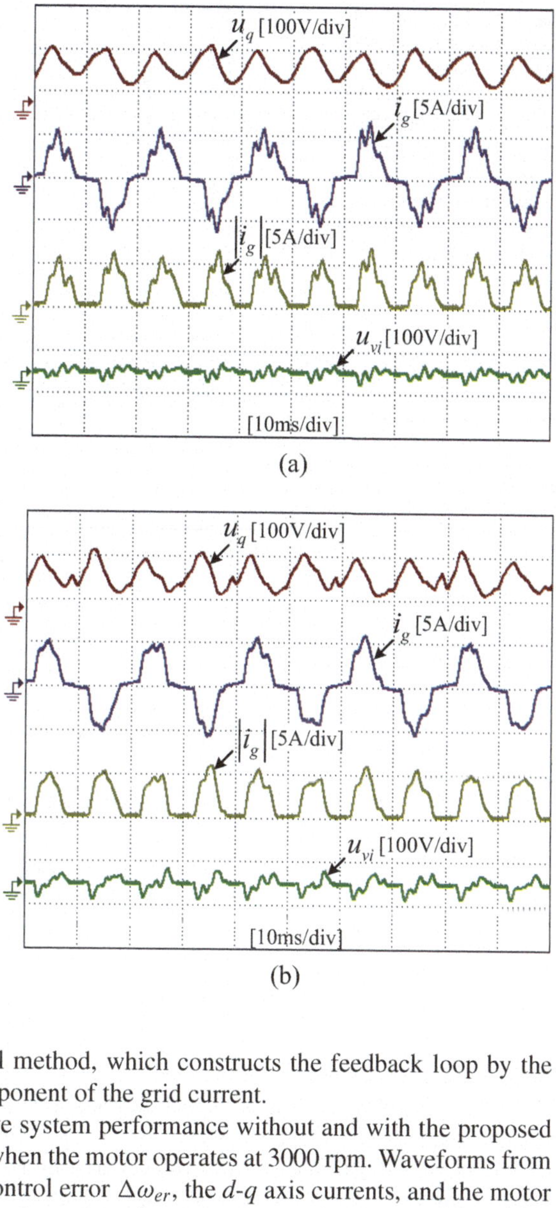

Fig. 4.13 Experimental waveforms of the q-axis voltage, grid current, absolute value of grid current, and voltage command generated by the proposed stabilization control method when the compressor operates at 3000 rpm. **a** Without the proposed stabilization control method. **b** With the proposed stabilization control method

proposed stabilization control method, which constructs the feedback loop by the feedback of the resonant component of the grid current.

Figure 4.14 shows the drive system performance without and with the proposed stabilization control method when the motor operates at 3000 rpm. Waveforms from top to bottom are the speed control error $\Delta\omega_{er}$, the d-q axis currents, and the motor current, respectively. The d-q axis currents fluctuate with the DC-link voltage fluctuation because of the insufficient voltage margin in the current control loop when the DC-link voltage fluctuates around the valley values. Hence the motor torque could not be maintained constant, which leads to the motor speed fluctuation. Meanwhile, the load torque of the single rotary compressor fluctuates with the rotor position from zero to the maximum value, and the fluctuation becomes more obvious as the motor speed increases. Hence the speed control error in Fig. 4.14a is the combination of the

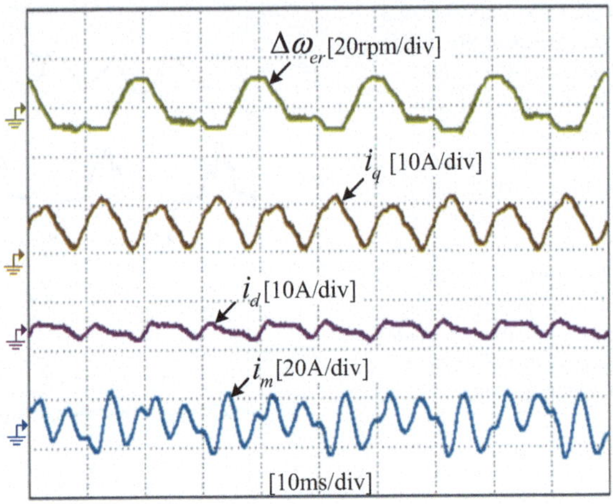

(a)

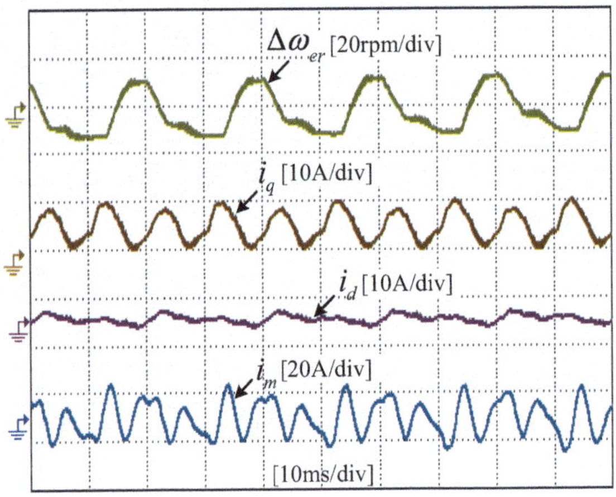

(b)

Fig. 4.14 Experimental waveforms of the speed control error $\Delta\omega_{er}$, the *d-q* axis currents, and the motor current when the compressor operates at 3000 rpm. **a** Without the proposed stabilization control method. **b** With the proposed stabilization control method

fluctuated DC-link voltage and the load torque. It can be seen that the speed error in Fig. 4.14b after applying the proposed stabilization control method is slightly larger than the one in Fig. 4.14a, because the frequency of the voltage command u_{vi} generated by the stabilization control method is around the resonant frequency (503 Hz) and the sum of u_{vi} is zero. The effect of the proposed stabilization control method on the motor current is little for the high impedance value of the stator windings around the resonant frequency (503 Hz). However, the severely fluctuated DC-link voltage leads to the unavoidable negative effect on the motor control performance, including the torque and speed fluctuation. Hence the reduced DC-link capacitance motor drives are suitable to the applications interested in the low cost and high reliability, which are more tolerant to the torque and speed ripple, such as the household appliances, the fan and pump applications.

4.5 Summary

The stability issue and the grid current harmonics issue are important concerns in the reduced DC-link capacitance motor drives. The impedance model of the reduced DC-link capacitance motor drive system is established in this chapter to analyze the drive system stability, which is realized by the Nyquist stability criterion of the cascade drive system. The grid input impedance model is used to analyze the harmonics of the grid current. Compared with the conventional DC-link voltage feedback based active damping method, the grid current feedback based active damping method could formulate the Nyquist plots of the drive system and suppress the sharply change value of the grid input impedance. Hence the stability of the drive system could be enhanced and the grid current performance could be improved simultaneously. Experimental results are provided to verify the effectiveness of the proposed method.

References

1. A. Awan, S. Pierfederici, B. Nahid-Mobarakeh, F. Meibody-Tabar, Active stabilization of a poorly damped input filter supplying a constant power load, in *IEEE Energy Conversion Congress and Exposition* (2009), pp. 2991–2997
2. Y. Feng, L. Mathe, K. Lu, F. Blaabjerg, W. Xiongfei, P. Davari, Analysis of harmonics suppression by active damping control on multi slim dc-link drives, in *IECON Annual Conference of the IEEE Industrial Electronics Society* (2016), pp. 5001–5006
3. X. Liu, A.J. Forsyth, A.M. Cross, Negative input-resistance compensator for a constant power load. IEEE Trans. Ind. Electron. **54**(6), 3188–3196 (2007)
4. Y.A.I. Mohamed, A.A.A. Radwan, T.K. Lee, Decoupled reference-voltage-based active DC-link stabilization for PMSM drives with tight-speed regulation. IEEE Trans. Ind. Electron. **59**(12), 4523–4536 (2012)
5. K. Pietilainen, L. Harnefors, A. Petersson, H.-P. Nee, DC-link stabilization and voltage sag ride-through of inverter drives. IEEE Trans. Ind. Electron. **53**(4), 1261–1268 (2006)
6. X. Song, S. Zheng, B. Han, C. Peng, X. Zhou, Active damping stabilization for high-speed BLDCM drive system based on band-pass filter. IEEE Trans. Power Electron. **32**(7), 5438–5449 (2017)
7. N. Zhao, G. Wang, D. Ding, G. Zhang, D.G. Xu, Impedance based stabilization control method for reduced DC-link capacitance IPMSM drives. IEEE Trans. Power Electron. **34**(10), 9879–9890 (2019)

Chapter 5
Analysis and Suppression of Beat Phenomenon

The beat phenomenon is an important issue in the reduced DC-link capacitance motor drive system and other power electronic converters, including the resonant switching converter, the inverter with low switching frequency, the distributed power systems, and the matrix converter [1–3]. The beat phenomenon is caused by the interaction of signals with different frequencies, which leads to the low frequency fluctuation. The power electronic converters usually own high gains in the low frequency domain. Hence the beat phenomenon could motivate the large ripple response, which increases the voltage and current stress, weakens the control performance, and causes additional noise [4–7].

As for the reduced DC-link capacitance motor drive system, a specific beat phenomenon is generated by the fluctuated DC-link voltage and the severely fluctuated load torque, which is seldom mentioned in previous researches. In this chapter, the analysis of the beat phenomenon is presented, which is caused by the interaction of two signals with specific frequencies (the frequency of the DC-link voltage and the frequency of the fluctuated load torque). The effect of the beat phenomenon on the grid current and the motor speed is investigated, which fluctuate with low-frequency oscillation. Hence the suppression method based on the power balancing controller and the fluctuated torque suppression method is applied to improve the control performance of the drive system.

5.1 Beat Phenomenon Simply Caused by DC-Link Voltage

According to the PWM theory, the three-phase voltages u_a, u_b and u_c can be simplified as

© The Author(s), under exclusive license to Springer Nature Singapore Pte Ltd. 2021

G. Wang et al., *Reduced DC-link Capacitance AC Motor Drives*,
https://doi.org/10.1007/978-981-15-8566-1_5

$$\begin{bmatrix} u_a(t) \\ u_b(t) \\ u_c(t) \end{bmatrix} = \frac{mu_{dc}}{2} \begin{bmatrix} \sin(\omega_e t + \varphi_e) \\ \sin\left(\omega_e t + \varphi_e - {}^{2\pi}/_3\right) \\ \sin\left(\omega_e t + \varphi_e + {}^{2\pi}/_3\right) \end{bmatrix} \tag{5.1}$$

where m and φ_e are the modulation index and the initial phase of the phase voltage, respectively. Ignoring harmonics of the DC-link voltage except the mean value and the fundamental one, the a-phase voltage $u_a(t)$ can be derived as

$$\begin{aligned} u_a(t) &= \frac{mu_{dc,0}}{2}\sin(\omega_e t + \varphi_e) + \frac{mu_{dc,1}}{2}\sin(\omega_e t + \varphi_e)\sin\left(2\omega_g t + \varphi_{dc,1}\right) \\ &= \frac{mu_{dc,0}}{2}\sin(\omega_e t + \varphi_e) - \frac{mu_{dc,1}}{4}\cos\left[\left(\omega_e + 2\omega_g\right)t + \varphi_e + \varphi_{dc,1}\right]. \\ &+ \frac{mu_{dc,1}}{4}\cos\left[\left(\omega_e - 2\omega_g\right)t + \varphi_e - \varphi_{dc,1}\right] \end{aligned} \tag{5.2}$$

It can be seen that the last term of $u_a(t)$ is a low frequency disturbance to the drive system when ω_e gets close to $2\omega_g$. The low frequency component of the phase voltage is defined as the beat phenomenon, which could stimulate the periodically fluctuation of the drive system, increase the power loss and reduce the stability.

The solution to this beat phenomenon is to apply the DC-link voltage feedforward method. The modified index modulation m_i can be written as

$$m_i = \frac{m}{u_{dc}}. \tag{5.3}$$

Substituting (5.3) into (5.1), the DC-link voltage fluctuation is compensated and the beat phenomenon will be eliminated. The type of the beat phenomenon is caused by the interaction of two signals (i.e., the fluctuated DC-link voltage and the motor phase voltage) with close frequencies, which could be compensated by the DC-link voltage feedforward method [8].

Another beat phenomenon in the reduced DC-link capacitance motor drive system will appear when the motor load is with severely periodic fluctuation. Hence the conventional compensation method in [8] is not suitable for this type of the beat phenomenon, which is seldom mentioned in previous researches.

5.2 Beat Phenomenon of Reduced DC-Link Capacitance IPMSM Drives

5.2.1 Effect of Fluctuated DC-Link Voltage on Motor Current

Due to the DC-link voltage fluctuation in the reduced DC-link capacitance motor drive system, the motor phase voltages are influenced by the combination of the conventional motor phase voltage and the fluctuated DC-link voltage. The three-phase voltages u_a, u_b and u_c are presented as

$$\begin{bmatrix} u_a(t) \\ u_b(t) \\ u_c(t) \end{bmatrix} = \frac{mu_{dc,0}}{2}\begin{bmatrix} \sin(\omega_e t + \varphi_e) \\ \sin\left(\omega_e t + \varphi_e - {2\pi}/{3}\right) \\ \sin\left(\omega_e t + \varphi_e + {2\pi}/{3}\right) \end{bmatrix} + \frac{mu_{dc,1}}{2}\sin\left(2\omega_g t + \varphi_{dc,1}\right)\begin{bmatrix} \sin(\omega_e t + \varphi_e) \\ \sin\left(\omega_e t + \varphi_e - {2\pi}/{3}\right) \\ \sin\left(\omega_e t + \varphi_e + {2\pi}/{3}\right) \end{bmatrix}. \tag{5.4}$$

Hence the motor voltages on the d-q synchronous frame u_d and u_q could be presented as

$$\begin{bmatrix} u_d \\ u_q \end{bmatrix} = \begin{bmatrix} \cos(\omega_e t) & \sin(\omega_e t) \\ -\sin(\omega_e t) & \cos(\omega_e t) \end{bmatrix}\begin{bmatrix} u_a \\ \frac{1}{\sqrt{3}}(u_a + 2u_b) \end{bmatrix}. \tag{5.5}$$

Substituting (5.4) to (5.5), u_d and u_q could be simplified as

$$\begin{bmatrix} u_d \\ u_q \end{bmatrix} = \begin{bmatrix} \left[\frac{mu_{dc,0}}{2} + \frac{mu_{dc,1}}{2}\sin\left(2\omega_g t + \varphi_{dc,1}\right)\right]\sin\varphi_e \\ -\left[\frac{mu_{dc,0}}{2} + \frac{mu_{dc,1}}{2}\sin\left(2\omega_g t + \varphi_{dc,1}\right)\right]\cos\varphi_e \end{bmatrix}. \tag{5.6}$$

It can be seen from (5.6) that the d-q axis voltages u_d and u_q are influenced by u_{dc} fluctuation. Taking the q-axis current control loop as an example, the modulation signal m and the actual q-axis current i_q could be presented as

$$\begin{aligned} \frac{di_q(t)}{dt} &= \frac{m(t)}{L_q}\left(K_u u_{dc} - \omega_e \psi_f - \omega_e L_d i_d\right) - \frac{R_s i_q(t)}{L_q} \\ \frac{dm(t)}{dt} &= K_{iqP}\left[\frac{di_q^*(t)}{dt} - \frac{di_q(t)}{dt}\right] + K_{iqI}\left[i_{qref}(t) - i_q(t)\right] \end{aligned} \tag{5.7}$$

where K_u is $-\cos(\varphi_e)/2$.

As for the severely fluctuated dc-link voltage in the reduced dc-link capacitance motor drive system, the voltage margin of the current control loop reduces as the back electromotive force increases. Hence i_q could be maintained constant by the regulation of PI controller when the motor operates at low speed, and it fluctuates with u_{dc} fluctuation when the motor speed increases, because the voltage margin of the current control loop is not able to regulate i_q to be constant [9, 10]. Hence the d-q axis currents i_d and i_q under u_{dc} fluctuation are presented as

$$\begin{bmatrix} i_d \\ i_q \end{bmatrix} = \begin{bmatrix} K_{d0} + K_{d1}\sin\left(2\omega_g t + \varphi_{dc,1}\right) \\ K_{q0} + K_{q1}\sin\left(2\omega_g t + \varphi_{dc,1}\right) \end{bmatrix} \tag{5.8}$$

where K_{d0}, K_{d1} K_{q0} and K_{q1} are the mean value and the fundamental component of the *d*-*q* axis currents, respectively.

5.2.2 Interaction Between DC-Link Voltage Fluctuation and Load Torque Fluctuation

Besides the fluctuated DC-link voltage, the mechanical load also affects the *d*-*q* axis currents. The load torque of the single rotary IPMSM compressor air conditioner fluctuates with the mechanical rotor angle, which is determined by the operational principle of the single rotary compressor. The maximum load torque per cycle is determined by the motor speed and the operation state of refrigeration cycle, which becomes larger as the motor speed increases. The load torque T_L can be presented as

$$T_L = T_{L,0} + \sum_{k=1}^{n} T_{L,k} \sin\left(k\omega_r t + \varphi_{r,k}\right) \tag{5.9}$$

where $T_{L,0}$, $T_{L,k}$ and $\varphi_{r,k}$ are the mean value, the amplitude, and the phase of *k*th harmonic component of the load torque, respectively.

Based on the above analysis, the motor electromagnetic torque T_e is determined by two factors in the reduced DC-link capacitance motor drive system, the load torque T_L and the fluctuated DC-link voltage. Different from the constant value in the conventional motor drive system, i_d and i_q contain two sinusoidal components ($2\omega_g$ and ω_r corresponding to the DC-link voltage fluctuation and the load torque fluctuation) in the steady state of the proposed motor drive system. The condition that causes the beat phenomenon is the interaction of two different frequency signals, hence the beat phenomenon will occur when the *d*-*q* axis currents contain the components with the frequency of DC-link voltage fluctuation and the load torque fluctuation, respectively.

Ignoring high order harmonics, i_d and i_q can be presented as

$$\begin{aligned} i_d &= i_{d,0} + a_{d,0} \sin\left(2\omega_g t + \varphi_{dc,1}\right) + b_{d,0} sin\left(\omega_r t + \varphi_{r,1}\right) \\ i_q &= i_{q,0} + a_{q,0} \sin\left(2\omega_g t + \varphi_{dc,1}\right) + b_{q,0} sin\left(\omega_r t + \varphi_{r,1}\right) \end{aligned} \tag{5.10}$$

where $a_{d,0}$, $a_{q,0}$, $b_{d,0}$ and $b_{q,0}$ are the amplitude of the component caused by the fluctuated DC-link voltage and the fluctuated load torque, respectively. As i_q is closely related to T_e, which determines the drive system performance. Hence i_q is used to analyze the beat phenomenon of the reduced DC-link capacitance motor drive system. Assuming ω_g is 50 Hz, i_q is shown in Fig. 5.1 to directly illustrate the beat phenomenon under different operation frequencies of ω_r (49 Hz, 50 Hz, and 51 Hz). It can be seen from Fig. 5.1a and c that i_q fluctuates with the low frequency when ω_r is around 50 Hz, and the low-frequency oscillation is with the same frequency

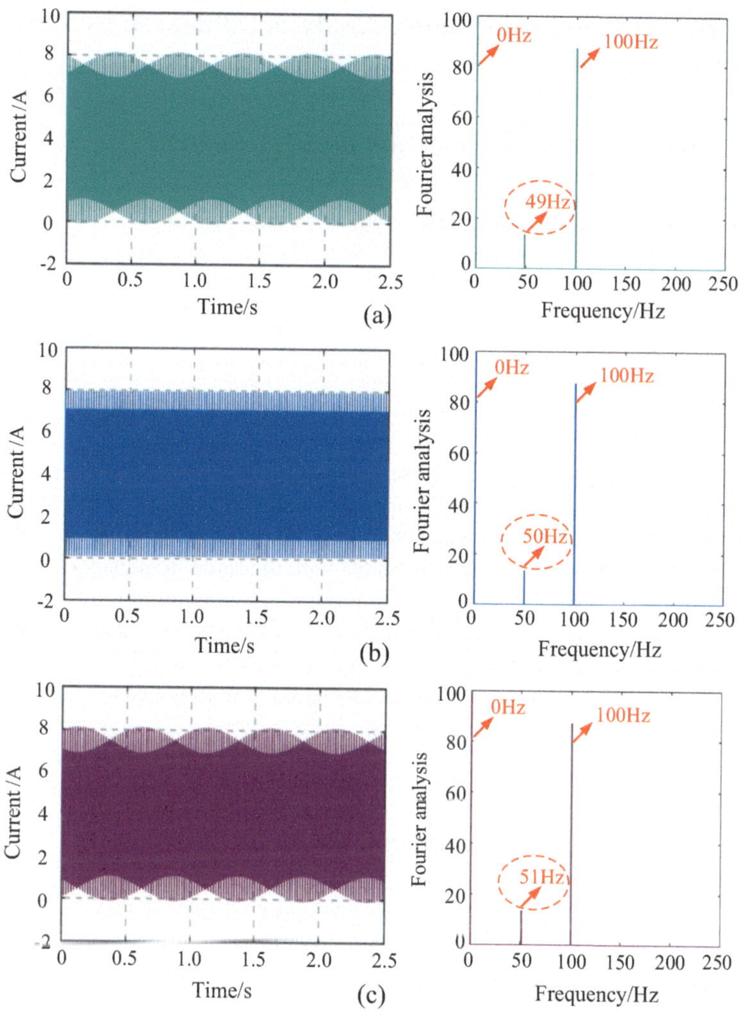

Fig. 5.1 Waveforms of i_q in the reduced DC-link capacitance motor drive system with different operation frequencies of ω_r. **a** 49 Hz. **b** 50 Hz. **c** 51 Hz

when ω_r is 49 and 51 Hz. In Fig. 5.1b, i_q is constant when ω_r is 50 Hz. Meanwhile, i_d is the same situation as i_q, which also fluctuates with the low frequency when ω_r is around 50 Hz. The frequency of the low-frequency oscillation increases as the frequency of motor speed diverges from 50 Hz. The low-frequency oscillation is the beat phenomenon of the drive system, which could stimulate the additional audible noise and possible resonance with the mechanical structure.

5.3 Drive System Performance Analysis Influenced by Beat Phenomenon

5.3.1 Effect of Beat Phenomenon on Grid Current

The grid input power P_g can be presented as

$$P_g = P_{inv} + P_{dc}. \tag{5.11}$$

P_{inv} could be presented as

$$P_{inv} = T_e \omega_r, \tag{5.12}$$

and P_{dc} is calculated by (2.5). It can be seen from (2.5) that the sum of P_{dc} per cycle is zero. Hence the grid input power P_g is equal to the inverter power P_{inv} when ignoring the instantaneous effect. P_g fluctuates with twice the frequency of the grid voltage. P_{inv} is closely related to T_L, which fluctuates with the same frequency of ω_r.

The grid input power P_g is also presented in (2.2). According to the analysis of (5.10) in Fig. 5.1, the beat phenomenon of P_g in (5.11) will be caused by the fluctuated inverter power. As the grid voltage u_g is determined by the power grid, the effect of the beat phenomenon on u_g is negligible. Hence the low-frequency oscillation of i_g will be caused by the beat phenomenon.

5.3.2 Effect of Beat Phenomenon on Motor Speed

Besides the grid current, the actual motor speed is also affected by the beat phenomenon, which fluctuates with the same low-frequency oscillation. The motor voltages in the synchronous frame u_d and u_q can be presented as

$$\begin{bmatrix} u_d \\ u_q \end{bmatrix} = \begin{bmatrix} R_s + sL_d & -\omega_e L_q \\ \omega_e L_d & R_s + sL_q \end{bmatrix} \begin{bmatrix} i_d \\ i_q \end{bmatrix} + \begin{bmatrix} 0 \\ \omega_e \psi_f \end{bmatrix}. \tag{5.13}$$

In order to evaluate the effect of the beat phenomenon on the drive system, the motor voltage, current and speed can be regarded as two parts: the mean value X_0 and the deviation to its mean value ΔX. Hence (5.13) can be presented as:

$$\begin{aligned} \begin{bmatrix} u_{d,0} + \Delta u_d \\ u_{q,0} + \Delta u_q \end{bmatrix} = \begin{bmatrix} R_s + sL_d & -(\omega_{e,0} + \Delta\omega_e) L_q \\ (\omega_{e,0} + \Delta\omega_e) L_d & R_s + sL_q \end{bmatrix} \cdot \\ \begin{bmatrix} i_{d,0} + \Delta i_d \\ i_{q,0} + \Delta i_q \end{bmatrix} + \begin{bmatrix} 0 \\ (\omega_{e,0} + \Delta\omega_e)\psi_f \end{bmatrix} \end{aligned} \tag{5.14}$$

The derivate operation s could be ignored in the steady state. The motor voltages are generated by the current controllers, which are presented as:

$$\Delta u_d = \left(K_{idP} + \frac{K_{idI}}{s}\right)\Delta i_d,\ \Delta u_q = \left(K_{iqP} + \frac{K_{iqI}}{s}\right)\Delta i_q \tag{5.15}$$

As the beat phenomenon is with the periodic fluctuation, the integral operation of Δi_d and Δi could be assumed as zero. Hence (5.14) can be simplified as:$_q$

$$\begin{aligned} L_q i_{q,0}\Delta\omega_e &= \left(R_s - K_{pd}\right)\Delta i_d - L_q\omega_{e,0}\Delta i_q \\ \left(L_d i_{d,0} + \psi_f\right)\Delta\omega_e &= \left(K_{pq} - R_s\right)\Delta i_q - L_d\omega_{e,0}\Delta i_d \end{aligned}. \tag{5.16}$$

It can be seen from (5.16) that $\Delta\omega_e$ is closely related to i_d and i_q, which are influenced by the disturbance with the frequencies corresponding to the DC-link voltage fluctuation and the load torque fluctuation ($2\omega_g$ and ω_r). The PI controller is applied in the speed control loop to maintain the motor speed constant. However, as for the component of $\Delta\omega_e$ with the frequency of $2\omega_g$, the speed PI controller is invalid when the voltage margin of the speed control loop is insufficient due to the increased back electromotive force, which leads to the unavoidable control errors with the frequency of $2\omega_g$. Meanwhile, as for the component of $\Delta\omega_e$ with the frequency of ω_r, the control error of the speed control loop is also caused by the periodically fluctuated load torque, because it is difficult to achieve a sufficient high bandwidth in actual applications to totally suppress the periodic control errors [11].

5.4 Beat Phenomenon Suppression Method

5.4.1 *Principle of Beat Phenomenon Suppression Method*

According to the above analysis, the proposed control method is shown in Fig. 5.2. The power balancing control method aims to shape P_{inv} to be synchronous with the

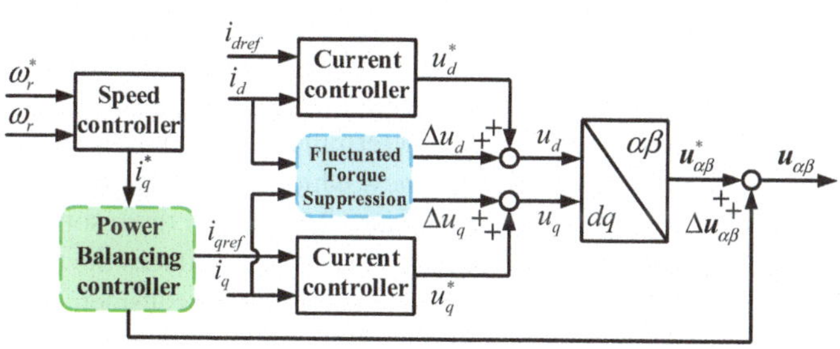

Fig. 5.2 Block diagram of power balancing and fluctuated torque suppression

fluctuated DC-link voltage. Hence the interaction of power signals in (5.10) will be changed, which could suppress the low-frequency oscillation of the grid current. The input of the drive system is the speed reference and load torque. The PI controllers are applied in the current and speed control loops to regulate the actual *d*-*q* axis currents and the motor speed. Due to the motor speed fluctuation caused by the beat phenomenon of i_d and i_q, the fluctuated torque suppression method is applied to eliminate the component of the *d*-*q* axis currents with the same frequency of ω_r. According to (5.10) and (5.16), the suppression of the signal with the same frequency of the fluctuated load torque (ω_r) could improve the performance of i_d and i_q. Hence the low-frequency oscillation of the motor speed could also be suppressed.

5.4.2 Beat Phenomenon Suppression of Grid Current

The block diagram of the power balancing control method is shown in Fig. 5.3. The power balancing controller is realized by the inverter power regulation, and voltage commands are applied to modify the voltage vector in the stationary frame directly. A resonant controller is applied to regulate the inverter power, because the inverter power fluctuates with twice the frequency of the grid voltage. Meanwhile, the *q*-axis current reference is modulated to be synchronous with the phase of the inverter power reference P_{ref}, which contributes to improving the performance of the power balancing strategy.

In Fig. 5.3, the phase of the grid voltage θ_g is detected by the PLL, which is used to generate the inverter power reference P_{ref} and modulate the *q*-axis current reference i_q^*. P_{ref} is determined by the load torque T_L and the actual motor speed ω_r. The DC-link capacitor power P_{dc} is ignored, because the sum of P_{dc} per cycle is zero and the frequency of the beat phenomenon shown in Fig. 5.1 is much lower than the one of P_{dc}. According to (5.11), P_{ref} is set as the inverter power reference and the actual inverter power P_{inv} is the feedback, which is obtained by the production of the motor voltage and current. Assuming the grid voltage is 50 Hz, the frequency of P_{ref} and P_{inv} is set as 100 Hz. Hence the resonant controller is applied to regulate the fluctuated power signals, which could be presented as

$$G_{PBC}(s) = K_{PBCR}\frac{2\omega_{pbc}s}{s^2 + 2\omega_{pbc}s + \omega_{pbc0}^2} \tag{5.17}$$

where K_{PBCR} is the resonant gain, which determines the amplitude of the resonant controller, and ω_{pbc0} is the resonant frequency, which is with the same frequency of P_{inv} (100 Hz), and ω_{pbc} is the cut-off frequency. The gain of the resonant controller increases as ω_{pbc} reduces. However, the decrease of ω_{pbc} reduces the robustness for the frequency shift of P_{inv}. Hence the compromise should be considered for the value determination of ω_{pbc}.

The output of the resonant controller P_{mod} can be transformed as the voltage command u_{mod}, which is presented as

$$u_{mod} = \frac{P_{mod}}{\sqrt{i_\alpha^2 + i_\beta^2}} = \frac{P_{mod}}{|i_{\alpha\beta}|}. \tag{5.18}$$

As for the normal motor drive system, the voltage command u_{mod} is a disturbance. Hence it is necessary to generate an optional way to reduce the impact on the drive system. The smallest amplitude of u_{mod} is applied, which is parallel to the current vector $\boldsymbol{i}_{\alpha\beta}$ in the stationary frame. The voltage command in the stationary frame can be presented as

$$\Delta u_\alpha = \frac{u_{mod} i_\alpha}{|\boldsymbol{i}_{\alpha\beta}|}, \Delta u_\beta = \frac{u_{mod} i_\beta}{|\boldsymbol{i}_{\alpha\beta}|}. \tag{5.19}$$

Meanwhile, the mean value of the DC-link voltage is applied as the voltage reference of the SVPWM module, which could avoid the unnecessary inverter saturation when the DC-link voltage fluctuates severely. As for the application of the power balancing strategy, the inverter output power will be synchronous with the grid input power, which could suppress the low-frequency oscillation of the grid current effectively.

5.4.3 Beat Phenomenon Suppression of Motor Speed

The performance of the grid current can be improved by the proposed power balancing control strategy. However, as analyzed previously, the motor speed fluctuates with the low frequency oscillation, which is caused by the beat phenomenon of i_d and i_q. The motor speed control performance is deeply affected by the severely fluctuated load torque T_L. Hence it is necessary to suppress the low frequency oscillation of the motor speed caused by the fluctuated torque.

According to (5.10), the beat phenomenon of i_d and i_q is generated by the interaction of two different frequency signals.

The periodic components of d-q axis currents with the frequency of ω_r could be extracted and eliminated to improve the motor speed performance. The block diagram is shown in Fig. 5.4. The closed loop is only sensitive to the fluctuated load torque, which is parallel to the normal operation of motor control. Hence harmonics of i_d and i_q caused by the fluctuated load torque could be effectively suppressed and the low frequency oscillation of the speed regulation will be eliminated.

The notch filters in Fig. 5.4 can be presented as

$$H(s) = \frac{a_{nf} s^2 + c_{nf} s + 1}{a_{nf} s^2 + b_{nf} s + 1}, \tag{5.20}$$

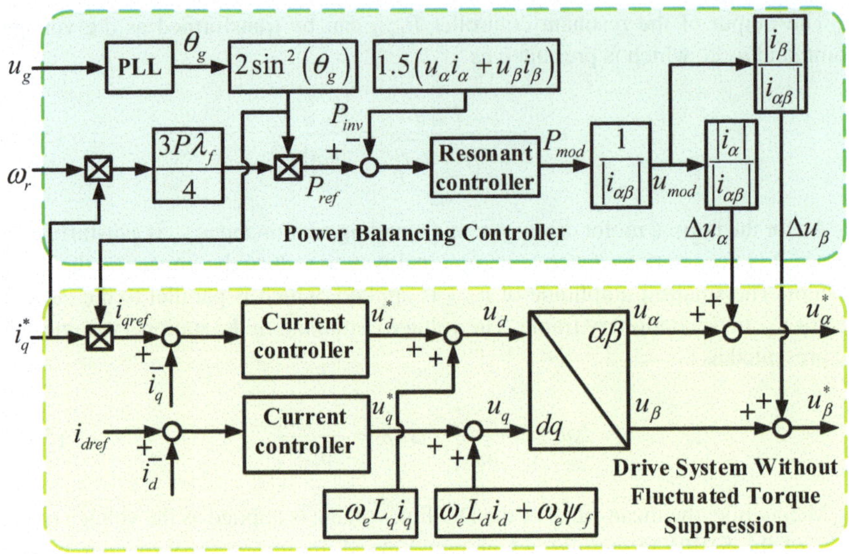

Fig. 5.3 Block diagram of power balancing controller

and

$$a_{nf} = \frac{1}{\omega_{nf0}^2}, b_{nf} = \frac{k_{nf1}}{\omega_{nf0}^2}, c_{nf} = \frac{k_{nf2}}{\omega_{nf0}^2} \tag{5.21}$$

where ω_{nf0}, k_{nf1} and k_{nf2} are the central frequency and coefficients of the notch filters, respectively. k_{nf1} determines the bandwidth of notch filter, and the attenuation is determined by k_{nf1} and k_{nf2}.

In this chapter, ω_{nf0} and k_{nf2} are set as 50 Hz and 0.001, and the determination of k_{nf1} (bandwidth of the notch filter) is an important issue in the actual realization, because the frequency of the beat phenomenon changes with the motor speed. The beat frequency increases as the frequency of motor speed diverges from 50 Hz, and the beat phenomenon will also be mitigated. Hence the determination of k_{nf1} should satisfy the requirement for the shift of the motor speed around 50 Hz when the beat phenomenon affects the drive system significantly, and the range of the frequency fluctuation is set as 4 Hz.

5.4.4 Experimental Results

As analyzed in the previous section, the beat phenomenon occurs when the motor speed gets close to the frequency of the grid voltage (50 Hz, 3000 rpm), hence the motor speed is set at 2960 rpm in experimental results. Figure 5.5 shows

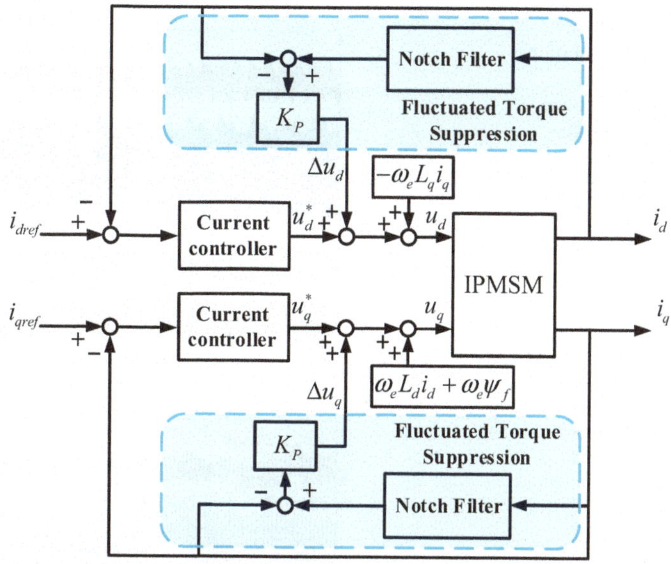

Fig. 5.4 Block diagram of fluctuated torque suppression

the system performance before and after applying the proposed beat phenomenon suppression method, and the bandwidth of the current controller is set as 200 Hz. Waveforms from top to bottom are the error of the motor speed control, the grid current, the inverter power, and the DC-link voltage, respectively. It can be seen from Fig. 5.5a that the motor speed and the grid current fluctuate with the low-frequency oscillation. The high frequency fluctuation (100 Hz) of the motor speed is caused by the fluctuated DC-link voltage, which is inevitable in the reduced DC-link capacitance drive system, and it is not considered in this chapter. The proposed method focuses on eliminating the low frequency oscillation generated by the beat phenomenon. The grid current fluctuation is obvious, and the motor speed also synchronously fluctuates with it (+16 rpm, −13 rpm), which weakens the drive system stability. The low frequency oscillation of the grid current is caused by the low frequency oscillation of the inverter power (+1150 W ~ +1400 W). In Fig. 5.5b, the power balancing controller is applied and the low frequency oscillation of the inverter power is suppressed effectively. Hence the performance of the grid current performance is also improved. However, as the power balancing controller only focuses on improving the performance of the grid current, the performance of the speed fluctuation is not improved in Fig. 5.5b (+17.5 rpm, −11.5 rpm) compared with the one in Fig. 5.5a. Combined with the power balancing controller, the fluctuated torque suppression method is applied in Fig. 5.5c. It can be seen that the low-frequency oscillation of the motor speed (+10 rpm, −10 rpm) decreases significantly and the drive system stability is enhanced. Hence the low-frequency oscillation caused by the beat phenomenon can be suppressed by the proposed method.

Fig. 5.5 Experimental waveforms of the error of the speed control, the grid current, the inverter power and the DC-link voltage. **a** Without the proposed method. **b** Only with the power balancing controller. **c** Combining the fluctuated torque suppression and the power balancing controller

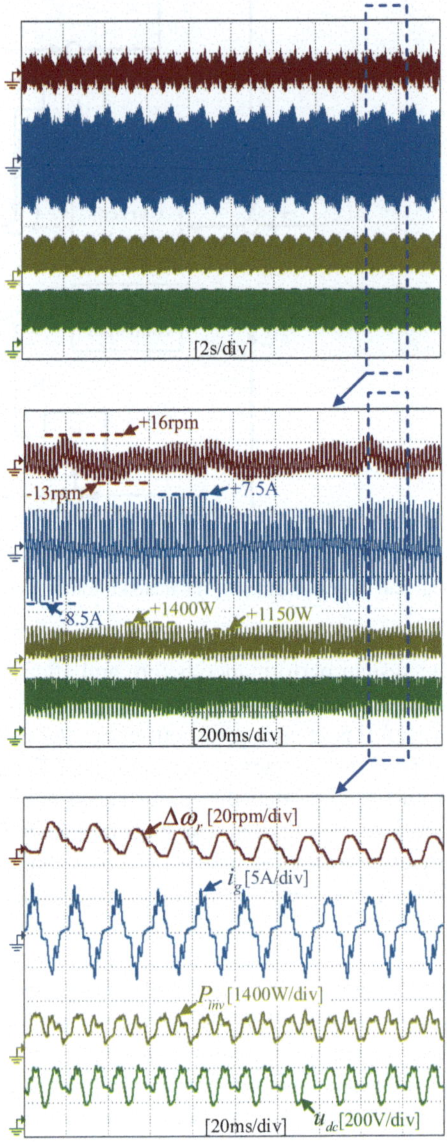

(a)

The motor phase current, the *d-q* axis currents are shown in Fig. 5.6. Meanwhile, the effect of the proposed method on the motor phase current could be obtained by the Fourier analysis in Fig. 5.6. Waveforms in Fig. 5.6a, b and c correspond to the system performance without the proposed method, only with the power balancing controller, and combining the fluctuated torque suppression method and the power balancing controller. It can be seen that the low-frequency oscillation of the *q*-axis current in

Fig. 5.5 (continued)

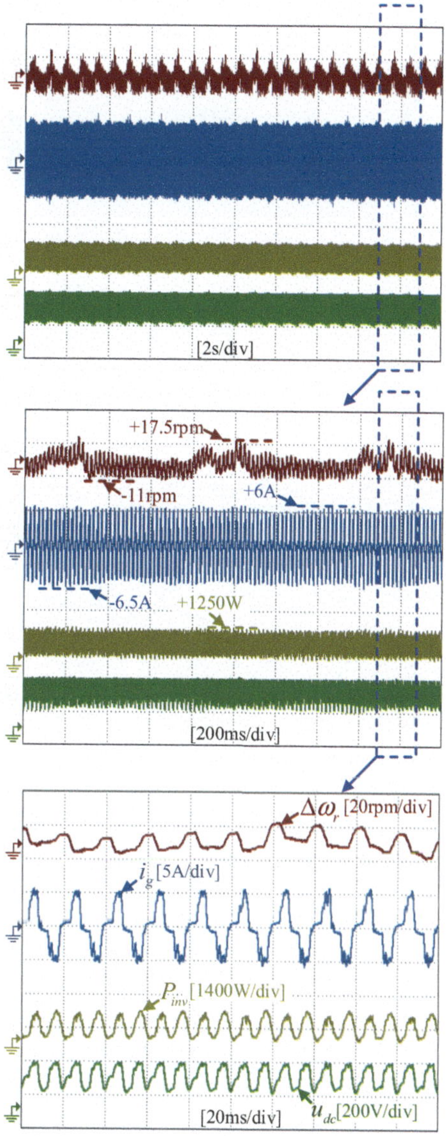

(b)

Fig. 5.6a (9A ~ 10.5A) is palpable, hence the motor speed fluctuation in Fig. 5.5a is serious. After applying the power balancing controller, the grid current fluctuation is suppressed, whereas the motor speed fluctuation is not improved as shown in Fig. 5.5b. It can be seen that the beat phenomenon of the q-axis current is also not effectively weakened in Fig. 5.6b (9.5A~10.5A). After applying the fluctuated torque suppression method, the low frequency fluctuation of the q-axis current in Fig. 5.6c

Fig. 5.5 (continued)

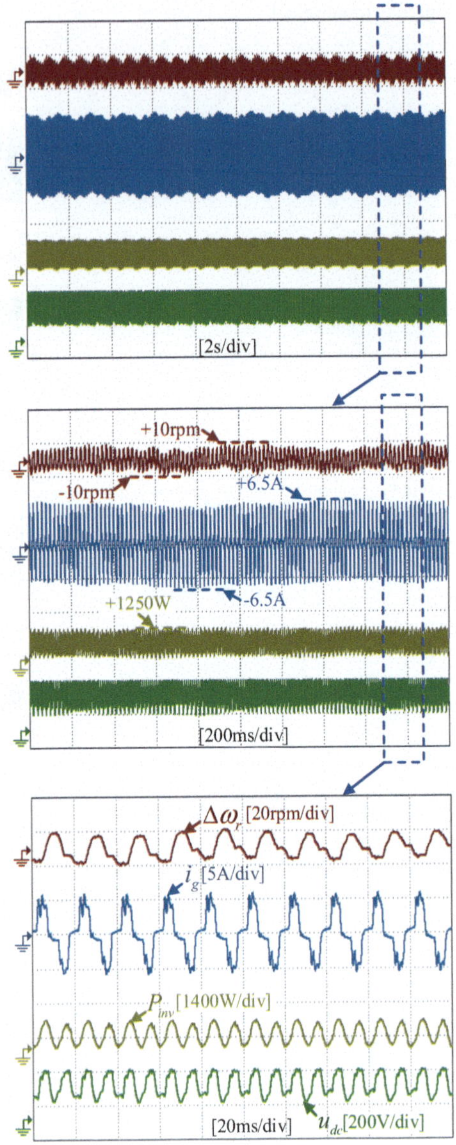

(c)

(10A~10.25A) is remarkably reduced compared with the ones in Fig. 5.6a and b. The suppression of the low-frequency oscillation of q-axis current could improve the motor speed performance, which is consistent with the experimental results in Figs. 5.5 and 5.6. The proposed power balancing controller and the fluctuated torque suppression affect the instantaneous performance of the motor phase current. As shown in Fig. 5.6, the main frequency of the motor current is 148 Hz, which is

Fig. 5.6 Experimental waveforms of the d-axis current, the q-axis current, and the motor phase current. **a** Without the proposed method. **b** Only with the power balancing controller. **c** Combining the fluctuated torque suppression and the power balancing controller

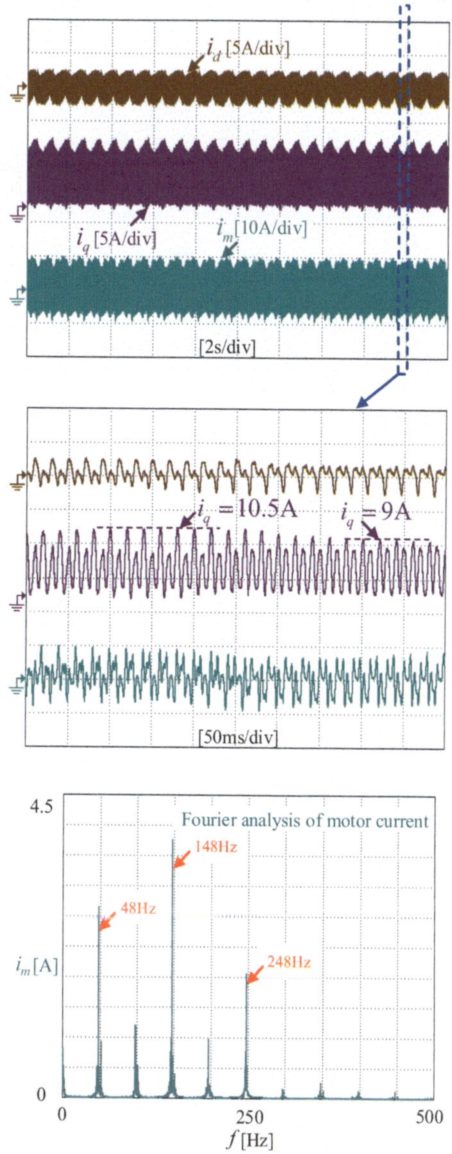

(a)

modulated by u_{dc}. Hence the additional harmonics are 48 and 248 Hz when the frequency of u_{dc} is 100 Hz. The application of the proposed method affects the percentage of harmonics distribution, whereas the additional DC component of the motor phase current is not generated by the proposed method.

Fig. 5.6 (continued)

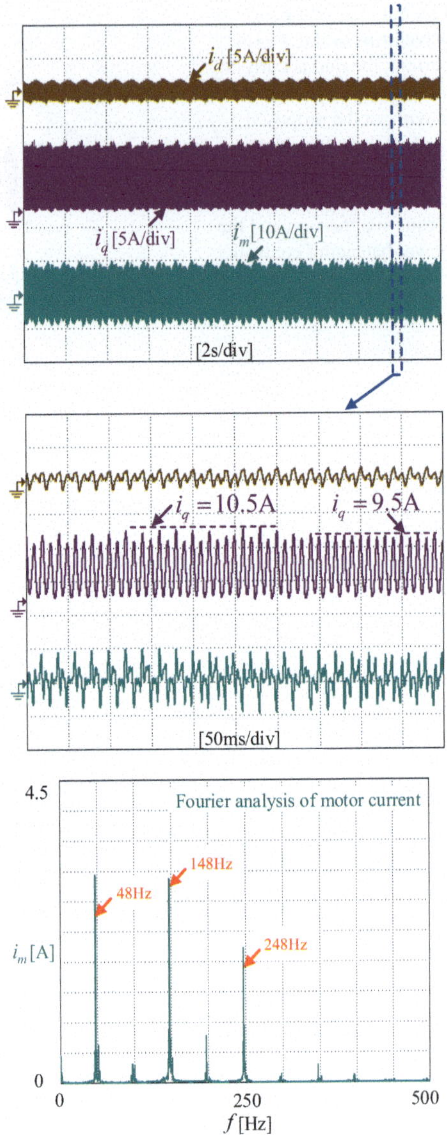

(b)

The effect of the fluctuated DC-link voltage on the motor current is affected by the parameters of the current controller. Hence experimental results are performed with a higher valley value in Fig. 5.7a varies from 0A~2.5A. The performance improvement of the current control loop applying a higher bandwidth current controller is limited, because the control loop is also affected by both the DC-link voltage fluctuation and the beat phenomenon. In Fig. 5.7b and c, the beat phenomenon of the *d-q* axis

Fig. 5.6 (continued)

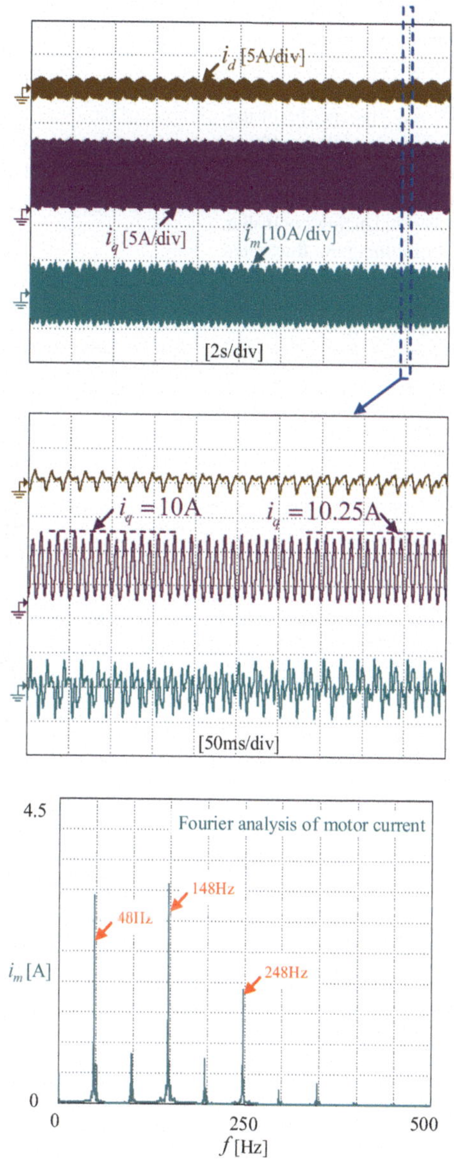

(c)

currents could be effectively regulated with the proposed beat suppression method. However, it can be seen that the noise of the phase current and the *d-q* axis currents increases in Fig. 5.7b and c, because the interaction of the normal current control and the proposed beat phenomenon method will generate the additional noise issue when applying a higher bandwidth current controller. Hence a relatively conservative

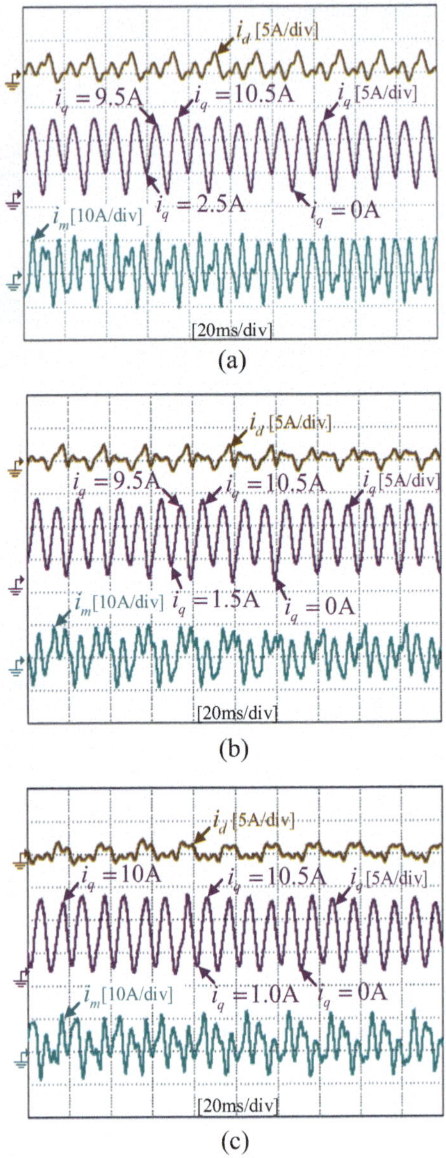

Fig. 5.7 Experimental waveforms of the d-axis current, the q-axis current, and the motor phase current when the bandwidth of the current controller is 400 Hz. **a** Without the proposed method. **b** Only with the power balancing controller. **c** Combining the fluctuated torque suppression and the power balancing controller

value of the bandwidth (200 Hz) is preferred to enhance the drive system stability, and the effectiveness of the proposed method could be verified by the experimental results with both high bandwidth (400 Hz) and low bandwidth (200 Hz) of the current controller.

5.5 Summary

A beat phenomenon suppression method for the reduced DC-link capacitance motor drive system is introduced in this chapter. The beat phenomenon is generated by the fluctuated DC-link voltage and the severely fluctuated load torque. The impact of the beat phenomenon on the drive system leads to the low-frequency oscillation of the grid current and the motor speed, which increases the audible noise and motivates the potential resonance with the mechanical system. The beat phenomenon of the grid current can be suppressed by the power balancing controller, and the fluctuated torque suppression method could improve the motor speed control performance. Experimental results with different bandwidth of the current controller are performed, and the relatively conservative bandwidth is preferred to prevent the drive system noise. Regardless of the current controller bandwidth, experimental results verify the effectiveness of the proposed method for the reduced DC-link capacitance motor drive system.

References

1. N. Zhao, G. Wang, B. Li, R. Zhang, D. Xu, Beat phenomenon suppression for reduced dc-link capacitance IPMSM drives with fluctuated load torque. IEEE Trans. Ind. Electron. **66**(11), 8334–8344 (2019)
2. X. Yue, F. Zhuo, S. Yang, Y. Pei, H. Yi, A matrix-based multifrequency output impedance model for beat frequency oscillation analysis in distributed power systems. IEEE J. Emerg. Sel. Top. Power Electron. **4**(1), 80–92 (2016)
3. S. Tian, F.C. Lee, Q. Li, A simplified equivalent circuit model of series resonant converter. IEEE Trans. Power Electron. **31**(5), 3922–3931 (2016)
4. J. Itoh, G.T. Chiang, K. Maki, Beatless synchronous PWM control for high-frequency single-pulse operation in a matrix converter. IEEE Trans. Power Electron. **28**(3), 1338–1347 (2013)
5. X. Yue, D. Boroyevich, F.C. Lee, F. Chen, R. Burgos, F. Zhuo, Beat frequency oscillation analysis for power electronic converters in dc nanogrid based on crossed frequency output impedance matrix model. IEEE Trans. Power Electron. **33**(4), 3052–3064 (2018)
6. Y. Miura, K. Inubushi, T. Yoshida, T. Fujikawa, T. Ise, Operation of modular matrix converter under close input and output frequency by using voltage space vector modulation, in *Annual Conference of the IEEE Industrial Electronics Society* (2015), pp. 5136–5141
7. Y. Iwaji, T. Sukegawa, T. Okuyama, T. Ikimi, M. Shigyo, M. Tobise, A new PWM method to reduce beat phenomenon in large-capacity inverters with low switching frequency. IEEE Trans. Ind. Appl. **35**(3), 606–612 (1999)
8. H. Ouyang, K. Zhang, P. Zhang, Y. Kang, J. Xiong, Repetitive compensation of fluctuating DC link voltage for railway traction drives. IEEE Trans. Power Electron. **26**(8), 2160–2171 (2011)
9. Y. Son, J. Ha, Direct power control of a three-phase inverter for grid input current shaping of a single-phase diode rectifier with a small DC-link capacitor. IEEE Trans. Power Electron. **30**(7), 3794–3803 (2015)
10. H. Lamsahel, P. Mutschler, Permanent magnet drives with reduced dc-link capacitor for home appliances, in *Annual Conference of IEEE Industrial Electronics* (2009), pp. 725–730
11. J. Gao, X. Wu, S. Huang, W. Zhang, L. Xiao, Torque ripple minimisation of permanent magnet synchronous motor using a new proportional resonant controller. IET Power Electron. **10**(2), 208–214 (2017)

Chapter 6
Flux-Weakening Control Method

Reduced DC-link capacitance permanent magnet synchronous motor (PMSM) drives have many advantages, such as longer lifetime, reduced cost and higher power density. However, due to the DC-link voltage fluctuation, the DC-link voltage utilization rate and the torque ripples are two major issues in flux-weakening operation region. Solutions for the torque ripples reduction and the DC-link voltage utilization improvement in flux-weakening region is important. Many flux-weakening control methods for motor drives with large-volume electrolytic capacitors have been proposed during the last decades [1–7]. However, the flux-weakening control for reduced DC-link capacitance motor drives should consider the fluctuation of DC-link voltage to ensure the operation performance [8].

6.1 Conventional Flux-Weakening Control

The voltage closed-loop flux-weakening scheme (VCFS) has been regarded as a practical method as shown in Fig. 6.1. In the conventional motor drive, the maximum voltage u_{smax} can be easily determined by the constant DC-link voltage and the modulation scheme. In fact, the maximum current i_{smax} should also be limited by the rated current. Therefore, the motor voltage and current are always subject to the following limits,

$$u_d^2 + u_q^2 \le u_{smax}^2 \tag{6.1}$$

$$i_d^2 + i_q^2 \le i_{smax}^2 \tag{6.2}$$

Combining (6.1) and (6.2), the voltage boundary can be denoted as

$$(i_d L_d + \psi_f)^2 + (i_q L_q)^2 \le \frac{u_{smax}^2}{\omega_e^2}. \tag{6.3}$$

© The Author(s), under exclusive license to Springer Nature Singapore Pte Ltd. 2021
G. Wang et al., *Reduced DC-link Capacitance AC Motor Drives*,
https://doi.org/10.1007/978-981-15-8566-1_6

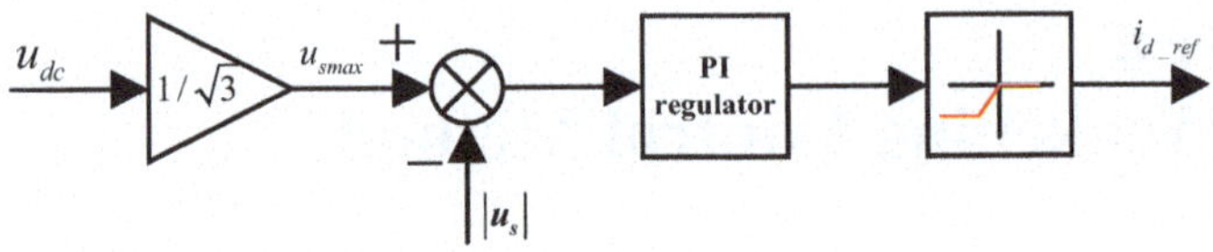

Fig. 6.1 Block diagram of the voltage closed-loop flux-weakening scheme

The voltage boundary expressed by (6.3) is an ellipse that is a function of ω_e. When the speed increases, the ellipse will shrink. If $|u_s|$ oversteps u_{smax}, the PI regulator will work to weaken the d-axis current.

However, u_{smax} is not easy to be determined due to the fluctuation of the DC-link voltage in the reduced DC-link capacitance drives.

6.2 Torque Ripple Analysis Caused by DC-Link Voltage Fluctuation

6.2.1 Introduction of Three-Phase Reduced DC-Link Capacitance PMSM Drives

The three-phase diode rectifier PMSM drive with reduced DC-link capacitance is shown in Fig. 6.2. It contains a three-phase diode rectifier bridge, film capacitors and an PMSM driven by a three-phase inverter. The volume of the DC-link capacitors has been largely reduced compared with the conventional drive.

Due to the use of film capacitors, the energy stored in the DC-link capacitors is relatively low. The energy coupling between the grid side and the motor side is aggravated. The energy required on the motor side will be higher than the energy stored in the film capacitors in most cases, so the energy will be transmitted from the grid side to the motor side directly. Generally, the frequency of the DC-link voltage is six-time that of the grid, as shown in Fig. 6.3.

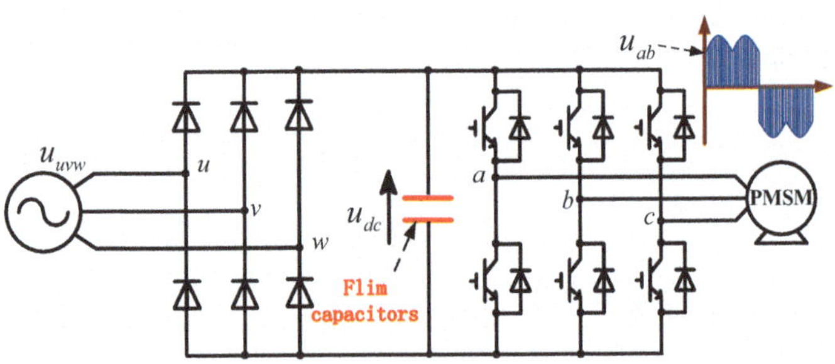

Fig. 6.2 Three-phase diode rectifier PMSM drive with reduced DC-link capacitance

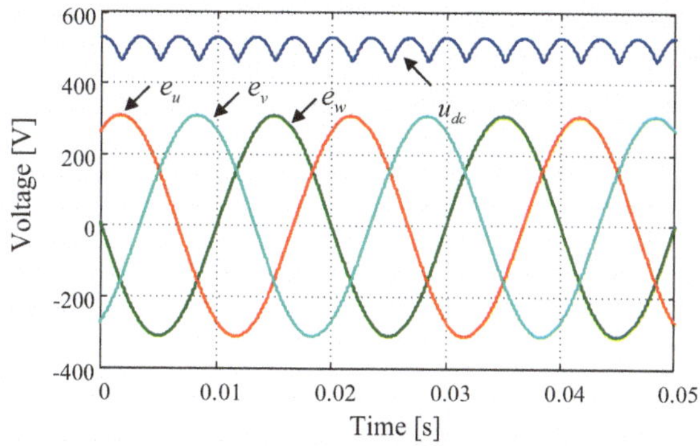

Fig. 6.3 Relationship between the DC-link and the grid-side phase voltages

The maximum and the minimum DC-link voltages are expressed in (6.4) and (6.5), respectively. The maximum DC-link voltage drop is about 72 V when the AC input of the inverter is 380-Vrms (50 Hz). The fluctuation of the DC-link voltage brings some difficulties in the control of motor drive, which will be discussed in the next section.

$$u_{dc\_max} = \sqrt{3}u_{u\_peak} \tag{6.4}$$

$$u_{dc\_min} = \sqrt{3}u_{u\_peak}\cos(\frac{\pi}{6}) \tag{6.5}$$

where u_{dc_max}, u_{dc_min}, u_{u_peak}, u_u, u_v, and u_w are the maximum DC-link voltage, the minimum DC-link voltage, the peak voltage of phase u, and the grid-side phase voltage of uvw, respectively.

6.2.2 Analysis of Influence on Stator Voltage

In the flux-weakening region, the reduction of DC-link capacitance will cause fluctuation in the stator voltages. Moreover, the overmodulation will aggravate this fluctuation. In this section, the stator voltage fluctuation will be analyzed. Further, the torque ripples caused by the DC-link fluctuation and the overmodulation will be discussed.

The Fourier expression of the DC-link voltage can be given as

$$u_{dc} = u_{dN}[1 - \sum_{n=1}^{\infty}\frac{2}{(6n)^2 - 1}cos(6n\omega_g t + \varphi_g)] \tag{6.6}$$

$$u_{dN} = \frac{3\sqrt{3}}{\pi}u_{u\_peak} \tag{6.7}$$

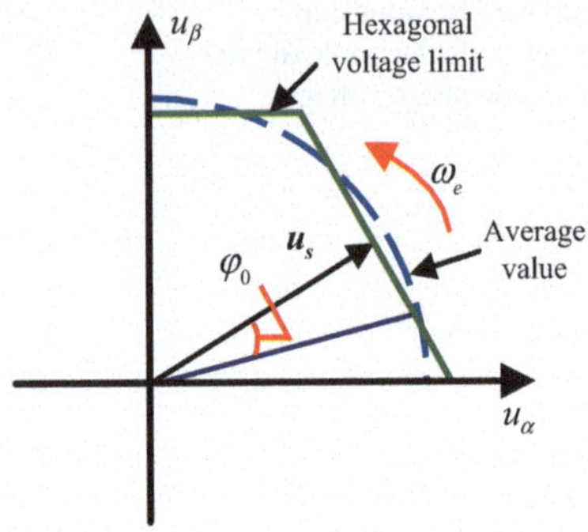

Fig. 6.4 Stator voltage vector and its average value in the minimum phase error overmodulation

where $n = 1, 2, 3 \ldots$, and u_{dN} is the DC component of u_{dc}.

Only considering the fundamental frequency, the DC-link voltage can be re-expressed as

$$u_{dc} = u_{dN}[1 - \frac{2}{35}cos(6\omega_g t + \varphi_g)]. \tag{6.8}$$

In this control system, applying the minimum phase error overmodulation in the SVPWM modulation strategy [9], the trajectory of the stator voltage vector u_s can be extended to the hexagon. Therefore, the average value of the stator voltage vector can be represented as the dash line in Fig. 6.4, assuming that the DC-link voltage is a constant value.

The stator voltage vector amplitude can be denoted as

$$|\mathbf{u_s}| = mu_{dc}[1 - A\sin(6\omega_e t + \varphi_o)], \tag{6.9}$$

$$m = \frac{\sqrt{2/\pi}}{\sqrt[4]{3}} \tag{6.10}$$

where u_s, $|u_s|$, A, ω_e and φ_o are the stator voltage vector, the amplitude of the stator voltage vector, the fluctuation coefficient of the stator voltage in overmodulation, the electrical speed, and the overmodulation initial angle, respectively.

Substituting (6.8) into (6.9), the amplitude of the stator voltage vector considering the fluctuated DC-link voltage and the overmodulation can be yielded,

$$|\mathbf{u_s}| = mu_{dN}[1 - \frac{2}{35}cos(6\omega_g t + \varphi_g)][1 - A\sin(6\omega_e t + \varphi_o)]. \tag{6.11}$$

In order to reveal the harmonic frequencies, (6.11) is further expressed as

$$
\begin{aligned}
|\mathbf{u}_\mathbf{s}| =& mu_{dN} - Amu_{dN}\sin(6\omega_e t + \varphi_o) \\
&-\frac{2}{35}mu_{dN}\cos(6\omega_g t + \varphi_g) \\
&-\frac{1}{35}Amu_{dN}sin[6(\omega_g - \omega_e)t + \varphi_g - \varphi_o] \\
&+\frac{1}{35}Amu_{dN}sin[6(\omega_g + \omega_e)t + \varphi_g + \varphi_o]
\end{aligned}
\quad (6.12)
$$

As shown in (6.12), in flux-weakening region, the fluctuated DC-link voltage and the overmodulation will lead to lots of harmonics. The frequency of the harmonics caused by the DC-link voltage fluctuation is $6\omega_g$, and the one regarding to overmodulation is $6\omega_e$. The fluctuation of DC-link voltage and the overmodulation will cause a low-frequency oscillation at the frequency of $6(\omega_g-\omega_e)$ and a high-frequency harmonic at the frequency of $6(\omega_g+ \omega_e)$. In fact, the high-frequency harmonic at $6(\omega_g+ \omega_e)$ can be ignored, due to the limited bandwidth of the current control loop.

In the *dq* synchronous rotating coordinate frame, the stator voltage vector can be expressed as

$$
\mathbf{u}_\mathbf{s} = |\mathbf{u}_\mathbf{s}|e^{j(\omega_e t+\varphi_u)} \quad (6.13)
$$

where φ_u is the angle between the *d*-axis and the stator voltage vector.

In the flux-weakening region, the stator voltage vector will remain between its maximum amplitude $|u_{s\_max}|$ and the minimum amplitude $|u_{s\_min}|$ in the second quadrant as shown in Fig. 6.5. The amplitude and the angle of stator voltage vector are determined by the electromagnetic torque ($T_1 < T_2$) and the motor operating frequency. The voltage hexagon rotates at the motor operating frequency and varies according to the DC-link voltage.

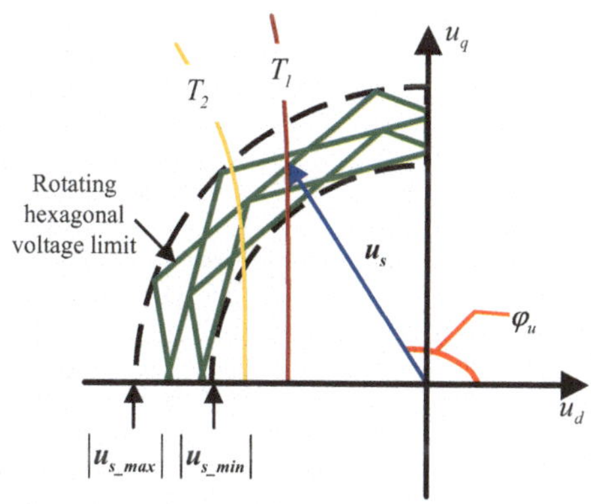

Fig. 6.5 Stator voltage vector in flux-weakening region

6.2.3 Analysis of Torque Ripple

At steady state, the relationship between the stator current and the stator voltage can be denoted as

$$i_d = \frac{u_q - \omega_e \psi_f}{\omega_e L_d}, \tag{6.14}$$

$$i_q = \frac{u_d}{-\omega_e L_q} \tag{6.15}$$

Combining (6.14) and (6.15), the electromagnetic torque can be expressed as a function of u_d and u_q,

$$T_e = 1.5P_n\left[\frac{u_d}{-\omega_e L_q}\psi_f + \left(L_d - L_q\right)\frac{u_d u_q}{-\omega_e^2 L_d L_q} + \frac{u_d \psi_f}{\omega_e L_d L_q}\right]. \tag{6.16}$$

According to (6.13), the expression can be obtained,

$$u_d u_q = |\mathbf{u_s}|^2 \cos(\varphi_u) sin(\varphi_u). \tag{6.17}$$

As shown in (6.12), $|u_s|$ has harmonic components at some specific frequencies. In fact, $|u_s|^2$ is very complicated, since it contains superposition of every harmonic component in $|u_s|$. However, the main harmonic components in $|u_s|^2$ are those multiplied by the relatively large DC part. Here, $|u_s|^2$ is simplified as

$$|\mathbf{u_s}|^2 = m u_{dN} |\mathbf{u_s}|. \tag{6.18}$$

The torque ripples at different harmonic frequencies can be derived as

$$\begin{aligned} T_{e1} &= -\frac{3}{2}P_n \frac{Am u_{dN}\psi_f \cos(\varphi_u)}{35\omega_e L_q} \\ &+ \frac{3}{2}P_n(L_d - L_q)[\frac{Am^2 u_{dN}^2 \cos(\varphi_u) sin(\varphi_u)}{-35\omega_e^2 L_d L_q} \\ &+ \frac{Am u_{dN}\psi_f \cos(\varphi_u)}{35\omega_e L_d L_q}] \end{aligned} \tag{6.19}$$

$$\begin{aligned} T_{e2} &= -\frac{3}{2}P_n \frac{Am u_{dN}\psi_f \cos(\varphi_u)}{\omega_e L_q} \\ &+ \frac{3}{2}P_n(L_d - L_q)[\frac{Am^2 u_{dN}^2 \cos(\varphi_u) sin(\varphi_u)}{-\omega_e^2 L_d L_q} \end{aligned}$$

$$+\frac{Amu_{dN}\psi_f\cos(\varphi_u)}{\omega_e L_d L_q}] \tag{6.20}$$

$$\begin{aligned}T_{e3} = &-\frac{3}{2}P_n\frac{2mu_{dN}\psi_f\cos(\varphi_u)}{35\omega_e L_q}\\ &+\frac{3}{2}P_n(L_d-L_q)[\frac{2m^2u_{dN}^2\cos(\varphi_u)sin(\varphi_u)}{-35\omega_e^2 L_d L_q}\\ &+\frac{2mu_{dN}\psi_f\cos(\varphi_u)}{35\omega_e L_d L_q}]\end{aligned} \tag{6.21}$$

where T_{e1}, T_{e2} and T_{e3} are the torque ripples at the frequencies of $6(\omega_g-\omega_e)$, $6\omega_e$ and $6\omega_g$, respectively.

As can be seen in (6.19), (6.20) and (6.21), the torque ripples are mainly determined by ω_e and φ_u. When the speed varies from 890 to 1210 r/min and φ_u varies in $[\pi/2, \pi]$, the quantitative analysis of torque ripples can be acquired as shown in Fig. 6.6 (motor parameters are given in Table 6.1).

Considering the parameters of the experimental PMSM listed in Table 6.1, the simulation results of the torque ripples without the flux-weakening control are shown in Fig. 6.7 when the speed is 980 r/min (49 Hz) and the load torque is 20 N·m. As can be seen in Fig. 6.7e, the torque ripples at the frequencies of 6 Hz, 306 Hz and 300 Hz are 0.04 N·m, 0.92 N·m and 1.12 N·m, respectively.

The stator voltage vector under this condition is given in Fig. 6.8. As can be seen, the amplitude of the stator voltage vector is 299 V, while the AC input of the inverter is 380-Vrms (50 Hz). Therefore, by using $\varphi_u = 0.61\pi$ rad and $\omega_e = 980$ r/min, the torque ripples can be obtained from Fig. 6.6, where at the frequencies of 6 Hz, 306 Hz and 300 Hz are about 0.05 N·m, 1.4 N·m and 1.7 N·m, respectively. The theoretical results are close to the simulation results described above, especially the low-frequency component.

Considering (6.19), (6.20), (6.21) and Fig. 6.6, the following conclusions in flux-weakening region can be obtained:

(1) As φ_u increases, the torque ripple will increase.
(2) The torque ripple at the frequency of $6(\omega_g-\omega_e)$ is relatively small compared with the ripples at the frequencies of $6\omega_e$ and $6\omega_g$ as shown in Fig. 6.6. However, the motor is more sensitive to the low-frequency oscillation in practical applications.

In order to reduce the torque ripple and increase the utilization rate of the DC-link voltage, the proposed adjustable maximum voltage based VCFS will be discussed in next section.

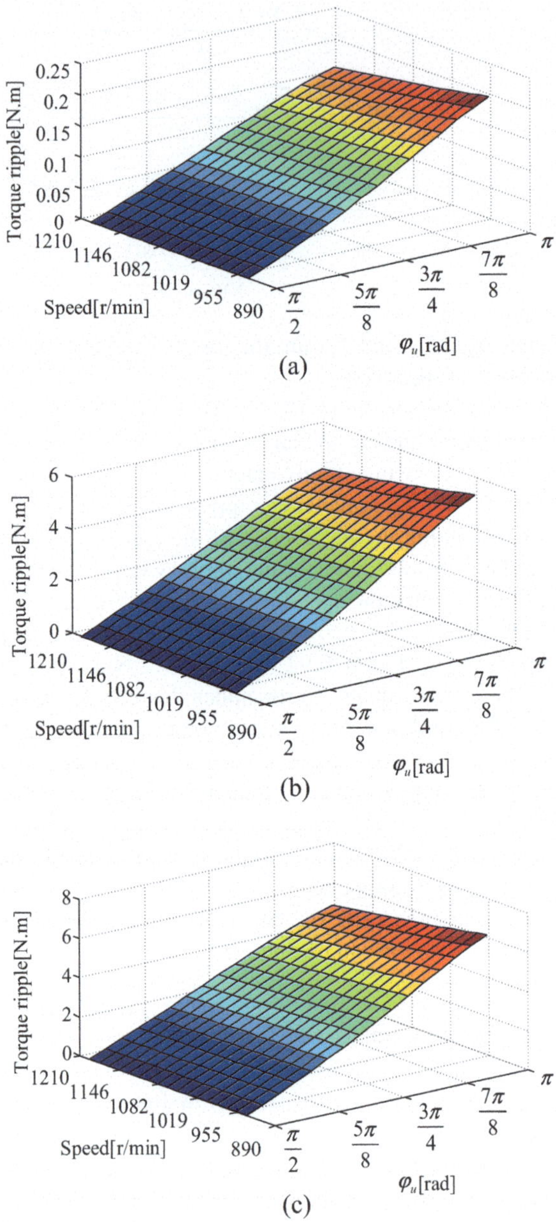

Fig. 6.6 Torque ripples at different frequencies. **a** Torque ripple at the frequency of $6(\omega_g$-$\omega_e)$. **b** Torque ripple at the frequency of $6\omega_e$. **c** Torque ripple at the frequency of $6\omega_g$

Table 6.1 PMSM parameters

Parameters	Value
Rated power	2.2 kW
Rated speed Rated current	1000 r/min 5.6 A
Stator resistance	2.75 Ω
d-axis inductance	35 mH
q-axis inductance	54 mH
Flux linkage of rotor	0.86 Wb
Number of pole pairs	3

6.3 Adjustable Maximum Voltage Based Flux-Weakening Control

6.3.1 Principle of the Control Method

The effective value of the DC-link voltage can be obtained as

$$u_{dc\_LPF} = \frac{u_{dc\_max} - u_{dc\_min}}{\sqrt{2}} + u_{dc\_min} \tag{6.22}$$

where u_{dc_LPF} is the DC-link voltage after the LPF.

Considering the relationship between u_{dc_min} and u_{dc_max}, u_{dc_min} can be expressed as

$$u_{dc\_min} = C_m u_{dc\_LPF}, \tag{6.23}$$

$$C_m = \frac{2\sqrt{3}}{2\sqrt{2}+2\sqrt{3} - \sqrt{6}} \tag{6.24}$$

where C_m is the ratio of u_{dc_min} to u_{dc_LPF}.

Using (6.23), the minimum DC-link voltage can be obtained as shown in Fig. 6.9.

In the whole flux-weakening region, the principle of the proposed adjustable maximum voltage based VCFS can be concluded as follows:

(1) Initially, u_{dc_min} is used to generate u_{smax}. Since the fluctuation component of the DC-link voltage is not used, the torque ripple can be eliminated effectively with the minimum time delay of the DC-link voltage measurement.
(2) As the increase of the motor speed, the amplitude of the current vector reference $|i_{s\_ref}|$ will increase to the current vector limit i_{s_lim} nearby i_{smax}. Then, $|i_{s\_ref}|$ is controlled around i_{s_lim}, and the output power of the inverter increases by extending u_{smax}.

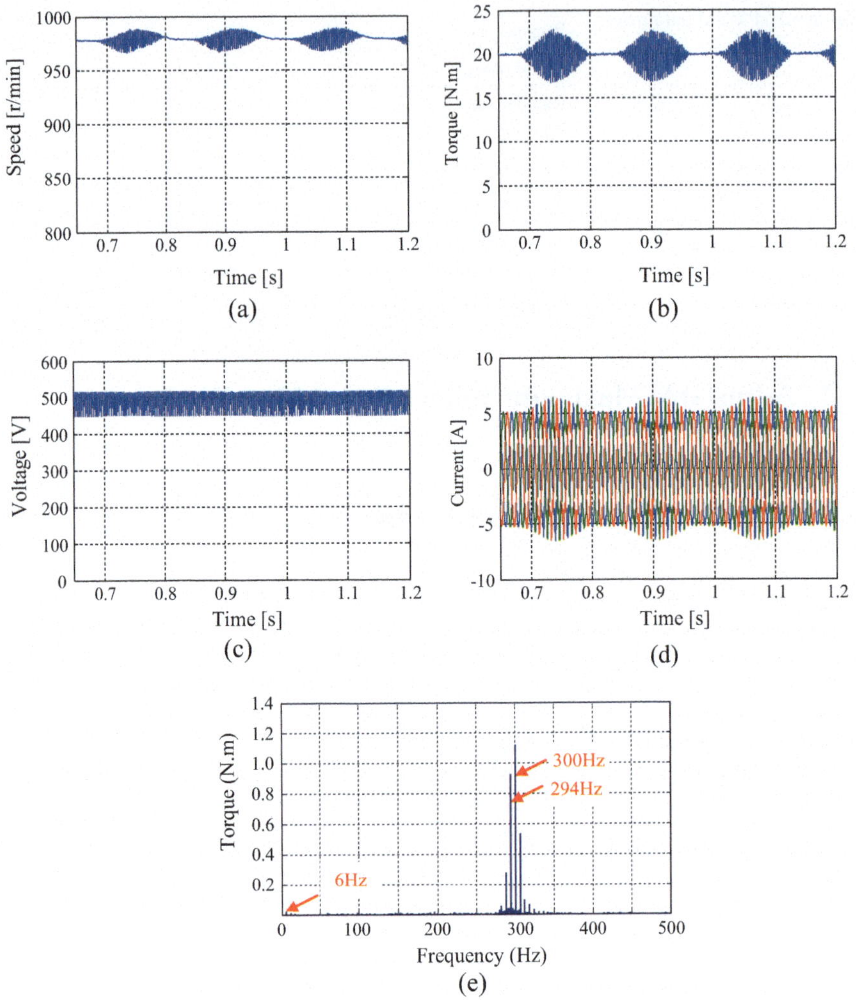

Fig. 6.7 Simulation results of torque ripples without the flux-weakening control at the motor frequency of 49 Hz. **a** Motor speed. **b** Electromagnetic torque. **c** DC-link voltage. **d** Stator currents. **e** FFT analysis of the electromagnetic torque

Fig. 6.8 Stator voltage vector under the simulation condition

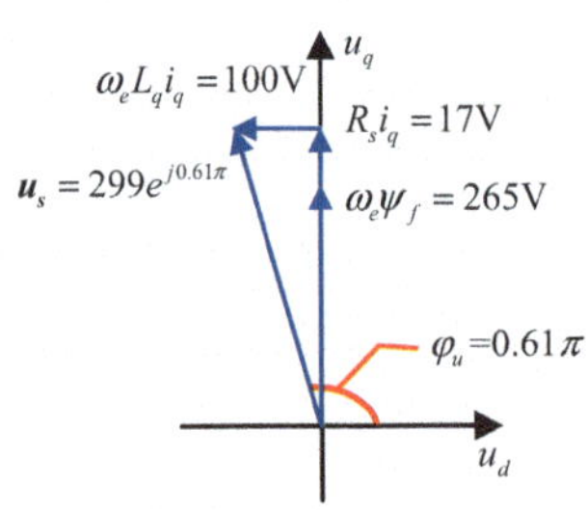

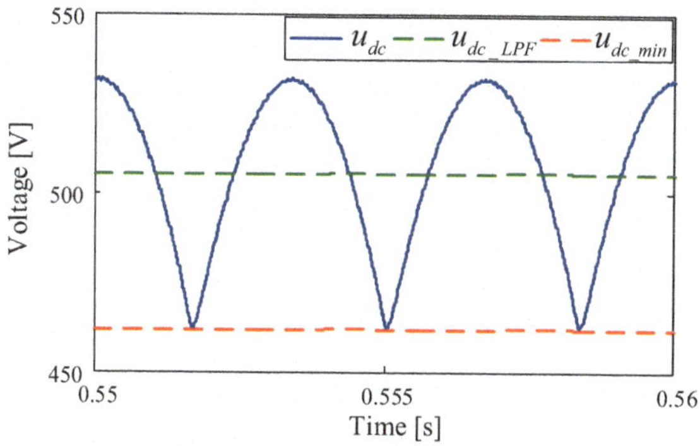

Fig. 6.9 The relationship among u_{dc}, u_{dc_LPF} and u_{dc_min}

6.3.2 Realization of the Control Method

The whole control system is shown in Fig. 6.10, and the proposed adjustable maximum voltage based VCFS is used in three-phase inverter with film capacitors. In the high-speed region, the position sensorless control based on back-electromotive force (EMF) is adopted.

In the flux-weakening controller, $|i_{s\_ref}|$ is used as a key factor to determine u_{smax}. As shown in Fig. 6.10, if $|i_{s\_ref}|$ is smaller than $i_{s\_lim\_1}$, which is the lower limit of the amplitude of the current vector, the minimum DC-link voltage control mode (Mode I) will be activated and u_{smax} is determined by u_{dc_min}.

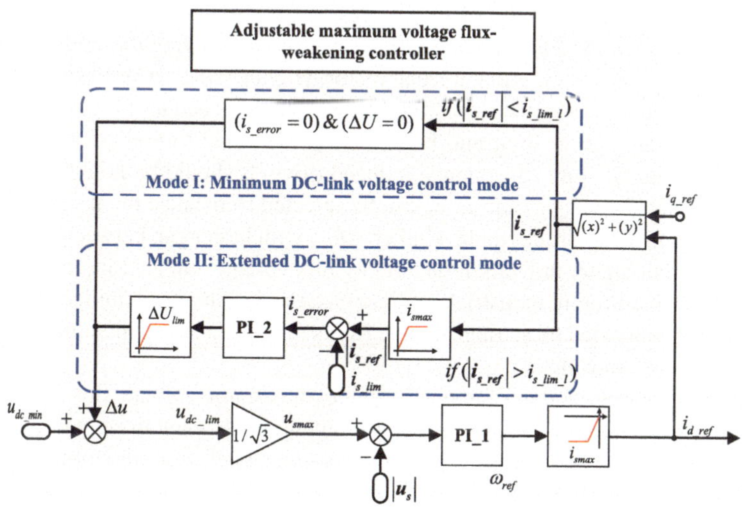

Fig. 6.10 Block diagram of the adjustable maximum voltage based VCFS in the three-phase inverter with film capacitors

The flux-weakening current can be expressed as

$$i_{d\_ref} = \left(k_{p1} + \frac{k_{i1}}{s}\right)\left(\frac{u_{dc\_min}}{\sqrt{3}} - |\mathbf{u}_s|\right) \tag{6.25}$$

where s is the Laplace operator, k_{p1} and k_{i1} are the coefficients of the PI_1 regulator, and i_{d_ref} is the instruction of d-axis current.

If $|i_{s\_ref}|$ is larger than $i_{s\_lim\_l}$, the extended DC-link voltage control mode (Mode II) will be activated. In the extended DC-link voltage control mode, $|i_{s\_ref}|$ is controlled around i_{s_lim} to adjust u_{smax} by the PI_2 regulator. When the output power needs to increase, $|i_{s\_ref}|$ will overstep i_{s_lim} and Δu will become a positive value to increase u_{dc_lim}.

$$\Delta u = \left(k_{p2} + \frac{k_{i2}}{s}\right)\left(\left|\mathbf{i_{s\_ref}}\right| - \mathbf{i_{s\_lim}}\right), \tag{6.26}$$

$$u_{dc\_lim} = \Delta u + u_{dc\_min} \tag{6.27}$$

where k_{p2} and k_{i2} are the coefficients in the PI_2 regulator, and Δu is the increment of u_{dc_lim}.

As a result, $|u_s|$ will increase to produce the expected output power by the PI_1 regulator. On the contrary, u_{dc_lim} will become lower when the expected output power decreases. Therefore, in the extended DC-link voltage control mode (Mode II), the output power is adjusted by the maximum voltage u_{smax}.

In the extended DC-link control mode (Mode II), i_{s_lim} is in the middle of $i_{s\_lim\_l}$ and i_{smax}, and the gap of the current limit between $i_{s\_lim\_l}$ and i_{smax} is set considering the torque ripples analyzed in the previous section, which means to ensure the effectiveness of the PI_2 regulator, this gap should be larger than the amplitude of the low-frequency oscillation current vector. Higher Δu_{lim}, which is the limit of Δu, will cause the increase of u_{dc_lim}, but it should not exceed u_{dc_max}. As analyzed, larger gap of the current limit and higher Δu_{lim} will lead to higher output power capability of the DC-link capacitance reduced drive. It should be noted that, Δu_{lim} and the gap between $i_{s\_lim\_l}$ and i_{s_lim} can be selected to balance the inverter output capability against the torque ripple. Higher Δu_{lim} with larger gap between $i_{s\_lim\_l}$ and i_{s_lim} will lead to higher utilization rate of DC-link voltage, but the torque ripple will also increase. In different industrial applications, the output power ability and torque ripple can be controlled by setting suitable limits of the adjustable maximum voltage flux-weakening controller.

The two control modes of the adjustable maximum voltage flux-weakening controller can be switched smoothly. As aforementioned, when the expected output power increases, the flux-weakening controller can switch from the minimum DC-link voltage control mode (Mode I) to the extended DC-link control mode (Mode II). Contrarily, the decrease of the expected outpower will reduce u_{dc_lim} until Δu is zero. In this condition, u_{dc_min} is used as u_{smax} again and $|i_{s\_ref}|$ will be lower than $i_{s\_lim\_l}$

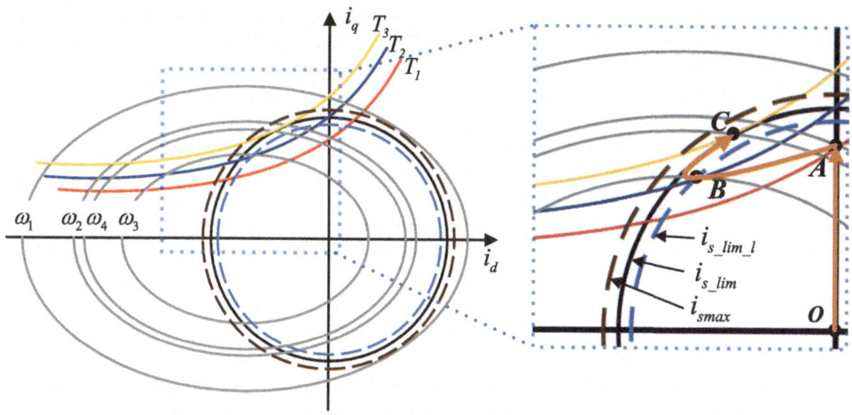

Fig. 6.11 Stator current vector trajectory in whole speed region

due to the losing effectiveness of the PI_2 regulator in the extended DC-link control mode. Therefore, the minimum DC-link voltage control mode will be activated.

In the extended DC-link voltage control mode (Mode II), the control system can be as stable as VCFS. When VCFS is used, the output power is increased by regulating the stator current. In the extended DC-link voltage control mode, the increase of output power is realized by increasing the stator voltage. In fact, from the energy perspective, these two flux-weakening strategies are the same.

6.3.3 Analysis of Stator Current Vector Trajectory

The system operation trajectory using the adjustable maximum voltage based VCFS is shown in Fig. 6.11. ω_1, ω_2, ω_3, ω_4 represent motor speed, where $\omega_1 < \omega_2 < \omega_3 < \omega_4$, and T_1, T_2, T_3 represent electromagnetic torque, where $T_1 < T_2 < T_3$. In fact, the voltage limit boundary is a hexagon and fluctuated due to the DC-link voltage fluctuation. For easy understanding, the ellipse is used to represent average voltage limit boundary. The brown dash line circle is current limit boundary, the blue dash line is $i_{s\_lim\_l}$, and the black solid circle is i_{s_lim}, which is located in the middle of the dash lines. Assuming that, the load torque increases when the speed gets higher. The current trajectory OA represents the accelerating process in no flux-weakening region. In the flux-weakening region, the current trajectory is shown as AB and BC. In the trajectory AB, u_{smax} is produced by the minimum DC-link voltage control mode (Mode I). As can be seen, the voltage limit boundary shrinks as the speed increases. The point B is the switching point of the minimum DC-link voltage control mode (Mode I) and the extended DC-link voltage control mode (Mode II).

As further increase of the motor speed, the current vector trajectory is represented by BC. During BC, the maximum voltage u_{smax} will increase since the fluctuation

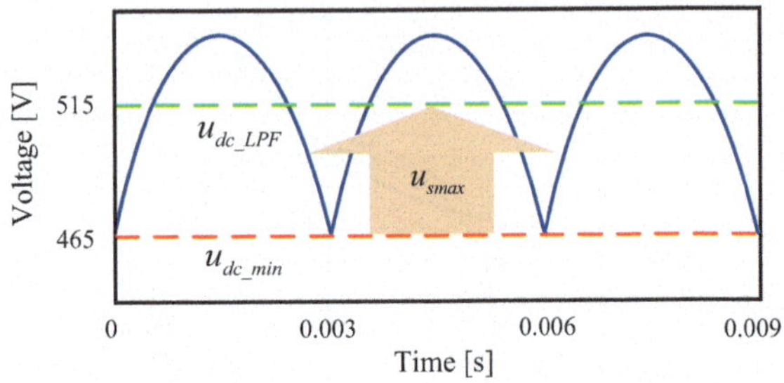

Fig. 6.12 Utilization of the DC-link voltage fluctuation component

component of the DC-link is used as shown in Fig. 6.12. Under this condition, the stator current vector i_s will always locate at the intersection of the voltage limit boundary and the stator current limit boundary, which is determined by the load torque. As the load torque increases, the current vector trajectory will move from B to C, and the ellipse will be enlarged since u_{smax}/ω_e is increased.

6.4 Power Loss Analysis of Flux-Weakening Control

In order to improve the torque performance, the flux-weakening method may increase the total losses of the PMSM and decrease the efficiency slightly in control mode I.

The total losses of PMSM can be denoted as [10]

$$P_{loss} = P_{Cu} + P_{Fe} + P_{str} + P_m \tag{6.28}$$

where P_{loss} is the total losses, P_{Cu} is the copper loss, P_{Fe} is the iron loss, P_{str} is the stray loss, and P_m is the mechanical loss.

The losses can be further expressed as

$$P_{Cu} = R_s\left(i_d^2 + i_q^2\right), \tag{6.29}$$

$$P_{Fe} = c_{Fe}\omega_e^{\beta} L_d^2\left(\left(\frac{\psi_f}{L_d} + i_d\right)^2 + \left(\frac{L_q^2}{L_d^2}\right)i_q^2\right), \tag{6.30}$$

$$P_{str} = c_{str}\omega_e^2\left(i_d^2 + i_q^2\right), \tag{6.31}$$

$$P_m = c_m\omega_e^2 \tag{6.32}$$

where c_{Fe}, c_{str} and c_m are the iron loss coefficient, the stray loss coefficient and the mechanical loss coefficient, respectively. R_s is the stator resistance, and $\beta = 1.5 \sim 1.6$.

Substituting (6.29), (6.30), (6.31) and (6.32) into (6.28), the total losses of the IPMSM can be obtained as

$$P_{loss} = a\left(i_d^2 + i_q^2\right) + b\left[\left(\frac{\psi_f}{L_d} + i_d\right)^2 + \left(\frac{L_q^2}{L_d^2}\right)i_q^2\right] + P_m \tag{6.33}$$

where

$$a = R_s + c_{str}\omega_e^2, \tag{6.34}$$

$$b = c_{Fe}\omega_e^{\beta} L_d^2. \tag{6.35}$$

Compared with the flux-weakening method proposed in [11], the fundamental component of the stator currents will increase since larger demagnetization current is needed at the same output condition in the minimum DC-link voltage control mode (Mode II). Therefore, the copper loss and stray loss will increase. Generally, the copper loss is the dominated component in the total losses while PMSM is operating around the based speed [12]. As a result, the proposed flux-weakening method has the limitation on the motor efficiency and the experimental results will be given in Sect. 6.5.

6.5 Experimental Results

The flux-weakening control algorithm is verified on a 2.2-kW IPMSM drive, supplied by a three-phase diode rectifier commercial drive with reduced DC-link capacitance as shown in Fig. 6.13. The AC input of the inverter is 380-Vrms (50 Hz). The PMSM parameters are listed in Table 6.1. A 50 μF film capacitor are used to replace the former 850 μF electrolytic capacitors in the DC side of the inverter. Another 2.2-kW IPMSM operating in the generation mode is used to provide the load torque. All the algorithms are implemented through ARM STM32F103, a 32-bit fixed-point microcontroller with 72 MHz maximum operating frequency. The inverter switching frequency is 6 kHz, the same as the current sampling frequency. In this study, k_{p1},

Fig. 6.13 Experimental setup of 2.2-kW IPMSM drive with film capacitors

k_{i1}, k_{p2} and k_{i2} are set as 2, 20, 0.2 and 0.8, respectively. The proportional and integral coefficients of the current regulator are set as 66 and 5000 respectively. And the proportional and integral coefficients of the speed regulator are set as 4 and 22, respectively.

The comparison of the torque ripple between the flux-weakening method in [11] and the adjustable maximum voltage based VCFS is shown in Fig. 6.14. The comparison is made in the minimum DC-link voltage control mode (Mode I). In Fig. 6.14a and b, the PMSM operates in 49 Hz (980 r/min). Using the method proposed by [11] as shown in Fig. 6.14a, the electromagnetic torques contains lots of harmonics especially at the frequency of 6, 294 and 300 Hz. The amplitude of the low-frequency oscillation at 6 Hz is about 0.14 N·m, and the range is corresponding to the analysis in section III. The amplitude of the high frequency fluctuations at 294 Hz and 300 Hz are 0.17 N·m and 0.26 N·m respectively, which are lower than the theoretical range for the limited bandwidth of the current loop. And the speed error is about 6 r/min. When the adjustable maximum voltage based VCFS is used, the torque ripple can be decreased effectively, and the speed error is smaller than 3 r/min, since the fluctuated component of the DC-link voltage is not used in this region, as shown in Fig. 6.14b. Similarly, when the PMSM operates in 51 Hz (1020 r/min), the torque ripples and speed error can also be reduced by comparing the waveforms shown in Fig. 6.14c and d. It should be noted that, when the speed increases, the amplitude of the low-frequency oscillation at $6(\omega_g - \omega_e)$ increases from 0.14 to 0.19 N·m by comparing the waveforms shown in Fig. 6.14a and c. This is due to the increase of φ_u, which verifies the analysis of the torque ripple changing tendency.

The experimental results at low demagnetization current are shown in Fig. 6.15. The demagnetization current changes from 0 to 5% to 15% of the current vector limit i_{s_lim}. The zoomed view while the demagnetization current is equal to 5% of i_{s_lim} as shown at the bottom of Fig. 6.15. As can be seen, the demagnetization current contains ripples at the frequency of 300 Hz and the amplitude of the ripples is about 1% of i_{s_lim}, which is due to the DC-link voltage measurement error. Due to the inertia of the PMSM, the speed is almost unaffected by the small current ripples.

The entire operation process of the adjustable maximum voltage based VCFS is shown in Fig. 6.16. $i_{s\_lim\_l}$ and i_{s_lim} are set to 95% (7.52 A) and 97% (7.67 A) of i_{smax}, respectively. Δu_{lim} is set to 9% (46 V) of u_{dc_LPF}. In region 1, $|i_{s\_ref}|$ is smaller than $i_{s\_lim\_l}$, and the minimum DC-link voltage control mode (Mode I) is activated, so i_d is reduced to increase the speed, and u_{dc_lim} is located in u_{dc_min}. The stator voltage is shown in Fig. 6.16b, where the blue dash line is the voltage hexagon considering u_{dc_min}, and the blue solid line is the voltage hexagon considering u_{dc_max}. The voltage vector is kept as the inscribe circle of the dash line. In region 2, the extended DC-link voltage control mode (Mode II) is activated, in which $|i_{s\_ref}|$ is greater than $i_{s\_lim\_l}$, so u_{smax} is increased to realize the improvement of the output power. The stator voltage can be seen in Fig. 6.16c, and it can extend beyond the dash line. It should be aware that, in region 2, the amplitude of i_a is a nearly constant, which means the amplitude of the stator current vector is limited at i_{s_lim}, and i_d is reallocated automatically when the speed and load torque furtherly increase. In region 3, $|i_{s\_ref}|$ increases until to i_{smax}, and u_{smax} reaches the predetermined value, indicating that the flux-weakening

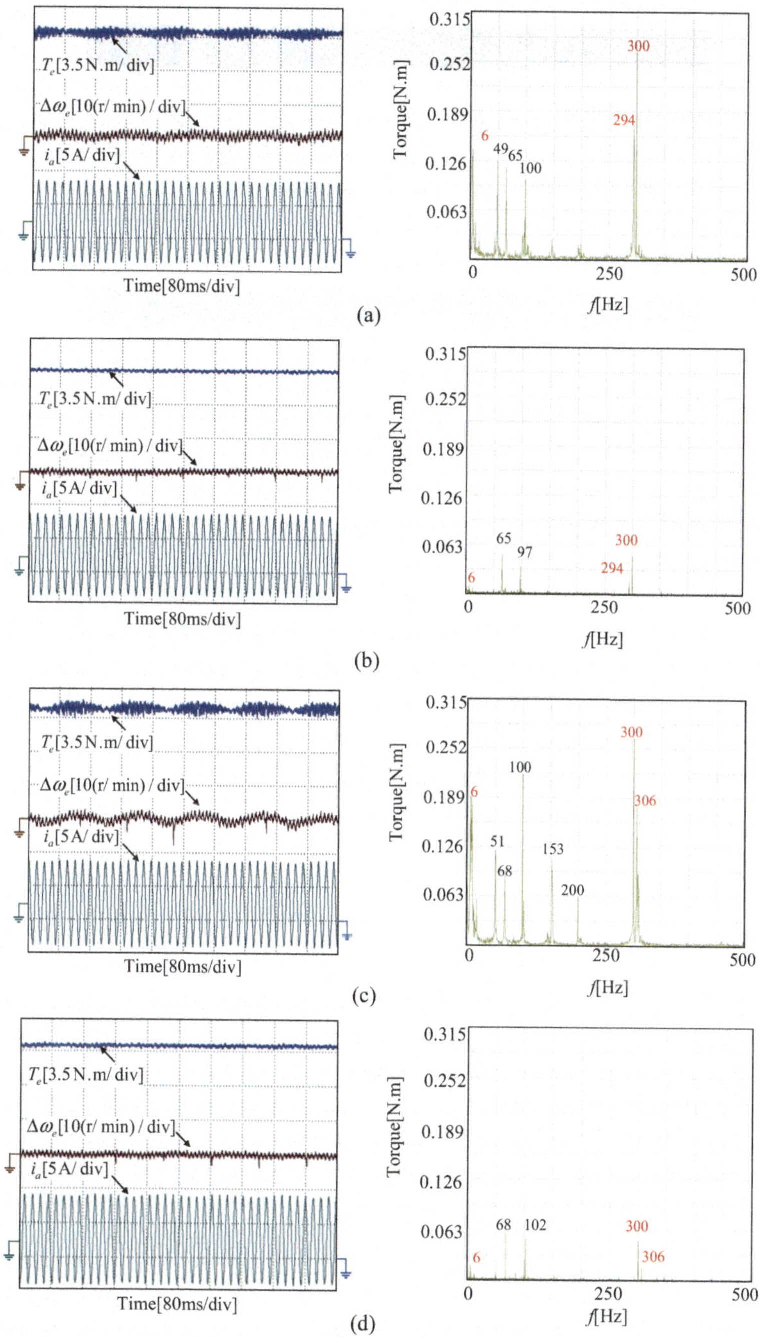

Fig. 6.14 Comparisons of torque ripple and speed error. **a** Method introduced in [21] at 49 Hz (980 r/min). **b** Adjustable maximum voltage based VCFS at 49 Hz. **c** Method introduced in [21] at 51 Hz (1020 r/min). **d** Adjustable maximum voltage based VCFS at 51 Hz

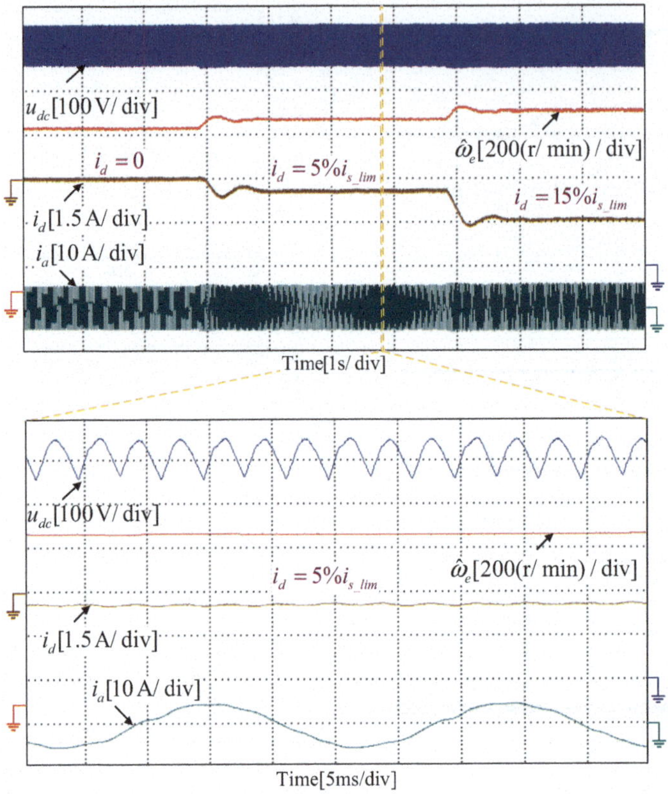

Fig. 6.15 Experimental results at low demagnetization current

controller is saturated. In region 4, the control ability can be reacquired when the speed decreases. As the motor speed decreases further, $|i_{s\_ref}|$ is smaller than $i_{s\_lim\_l}$, and the IPMSM operates in region 5.

In the flux-weakening region, the switching between two control modes will cause little affect to the speed, as shown in Fig. 6.17. In the accelerating process, the motor speed changes from 0 to 1150 r/min. When the control strategy varies from the minimum DC-link voltage control mode (Mode I) to the extended DC-link voltage control mode (Mode II), the speed is relatively smooth, as shown in the yellow region. When the IPMSM slows down to 1050 r/min, corresponding to the minimum DC-link voltage control mode (Mode I), the speed is not affected in decelerating process, as shown in the blue region.

To demonstrate the speed control performance, the experimental result with ramp speed change in two flux-weakening control modes is shown in Fig. 6.18 where the load torque increases with the motor speed. The motor speed changes between 940, 1100 and 1180 r/min, corresponding to no flux-weakening region, the minimum DC-link voltage control mode (Mode I) and the extended DC-link voltage control mode

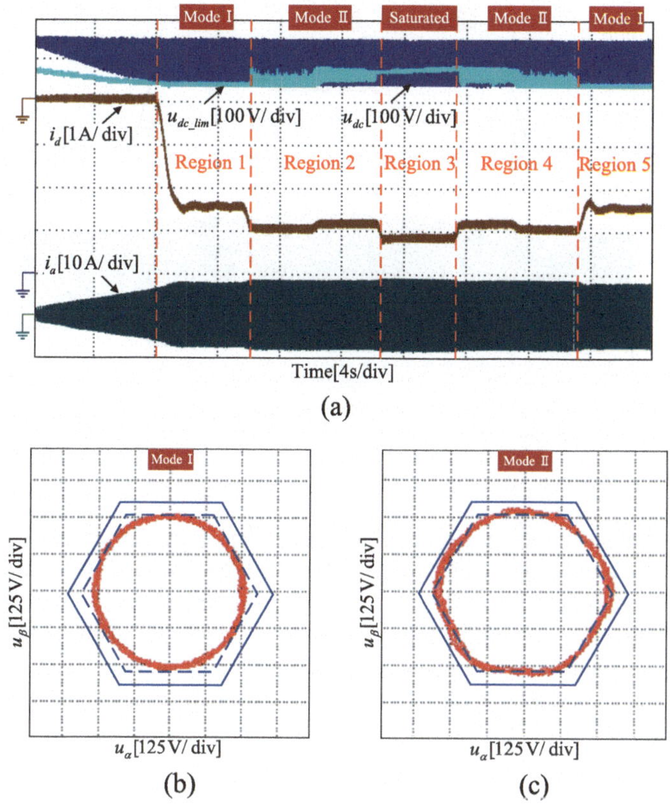

Fig. 6.16 Operation process of the adjustable maximum voltage based VCFS. **a** DC-link voltage and stator current in different operation mode. **b** Voltage vector trajectory in Mode I. **c** Voltage vector trajectory in Mode II

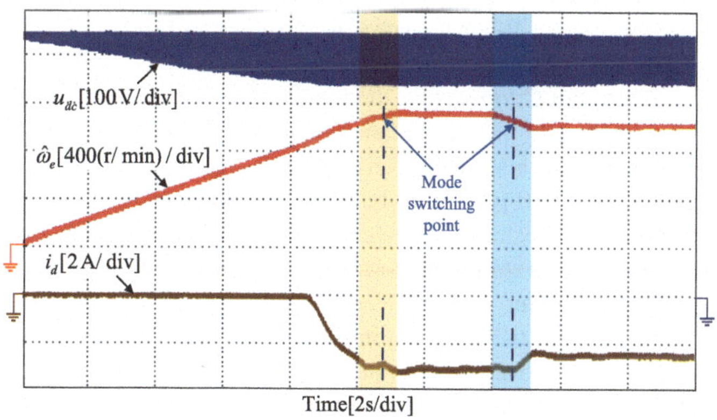

Fig. 6.17 Flux-weakening control mode change in accelerating and decelerating process (0 r/min–1150 r/min–1050 r/min)

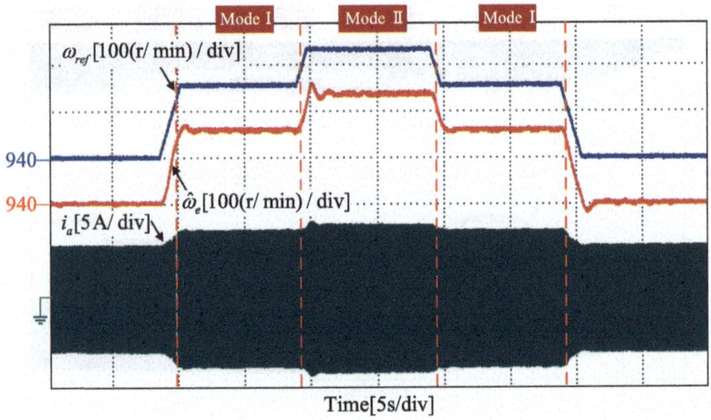

Fig. 6.18 Experimental result of the speed tracking performance (940 r/min–1100 r/min–1180 r/min)

(Mode II). As can be seen, the estimated speed can track the reference speed well in both two flux-weakening control modes. The estimated speed error is less than 10 r/min corresponding to 1% of the speed reference in the flux-weakening region.

By changing Δu_{lim} and the gap between $i_{s\_lim\_l}$ and i_{s_lim}, different maximum output stator voltages can be obtained as shown in Fig. 6.19. $i_{s\_lim\_l}$ and i_{s_lim} are set to 85% (6.72 A) and 92.5% (7.32 A) of i_{smax}, respectively. In Fig. 6.19a, b and c, Δu_{lim} is set to 9% (46 V), 12% (61 V) and 14% (71 V) of u_{dc_LPF}, respectively. As the increase of Δu_{lim}, the maximum output stator voltage can be extended. When Δu_{lim} is set to 14% of u_{dc_LPF}, the trajectory of the voltage vector can be extended to the inscribe circle of the hexagon determined by u_{dc_max}.

Figure 6.20 shows the current vector trajectory in the *dq* synchronous rotating coordinate frame. The red dash circle is i_{s_lim}, which is set to 97% (7.67 A) of i_{smax}. As can be seen, in the extended DC-link voltage control mode (Mode II), the current trajectory is settled in i_{s_lim}. When the speed increases from 1150 to 1170 r/min, u_{smax} is extended to improve the output power of the inverter. The current vector moves from the point B to point C to increase the load torque from 22 to 23 N·m.

The reduction of the DC-link capacitance can improve the grid side power quality dramatically. In this study, a 5.5-kW inverter with different DC-link capacitance (50 μF film capacitors and 850 μF electrolytic capacitors) is tested, respectively. The grid side power quality is tested by a power analyzer (WT1800). When the operating speed of the motor is 1200 r/min, the corresponding motor output power is 3.3 kW, and the THD of the inverter with reduced capacitance can be reduced to 38% by contrast, as shown in Fig. 6.21a and b. In the same output condition, the grid side power factor can be improved to one and half times approximately, as shown in Fig. 6.21c and d. It should also be noted that, the grid side power quality has been improved greatly along with the increase of the output power in the inverter with reduced DC-link capacitance.

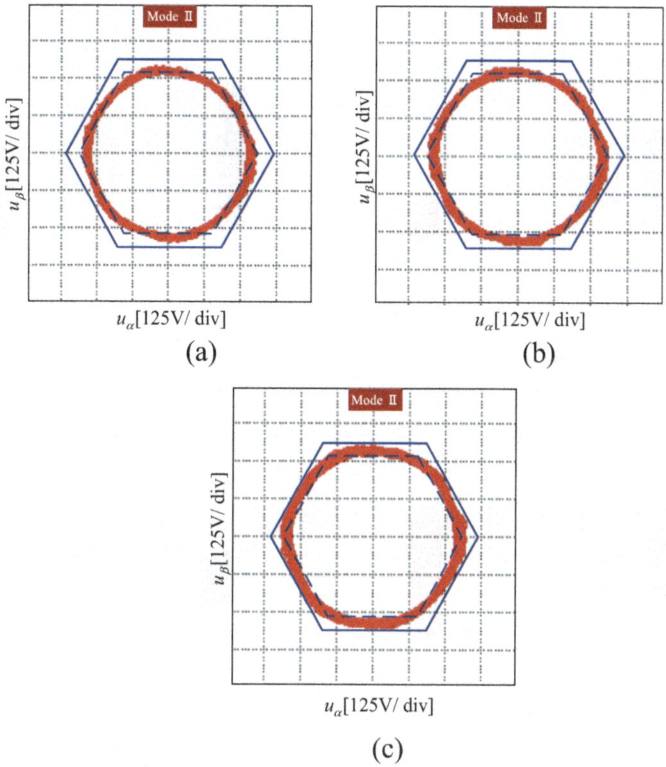

Fig. 6.19 Voltage vector trajectory with different limit in Mode II. **a** Δu_{lim} is set to 9% (46 V). **b** Δu_{lim} is set to 12% (61 V). **c** Δu_{lim} is set to 14% (71 V)

Fig. 6.20 Current vector trajectory when the speed and load torque increase at the same time in Mode II

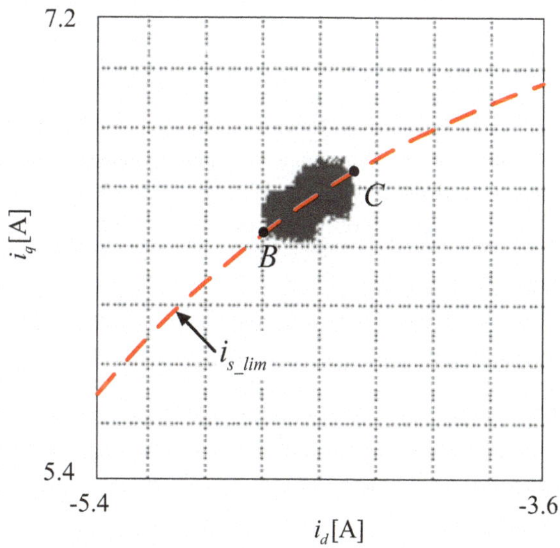

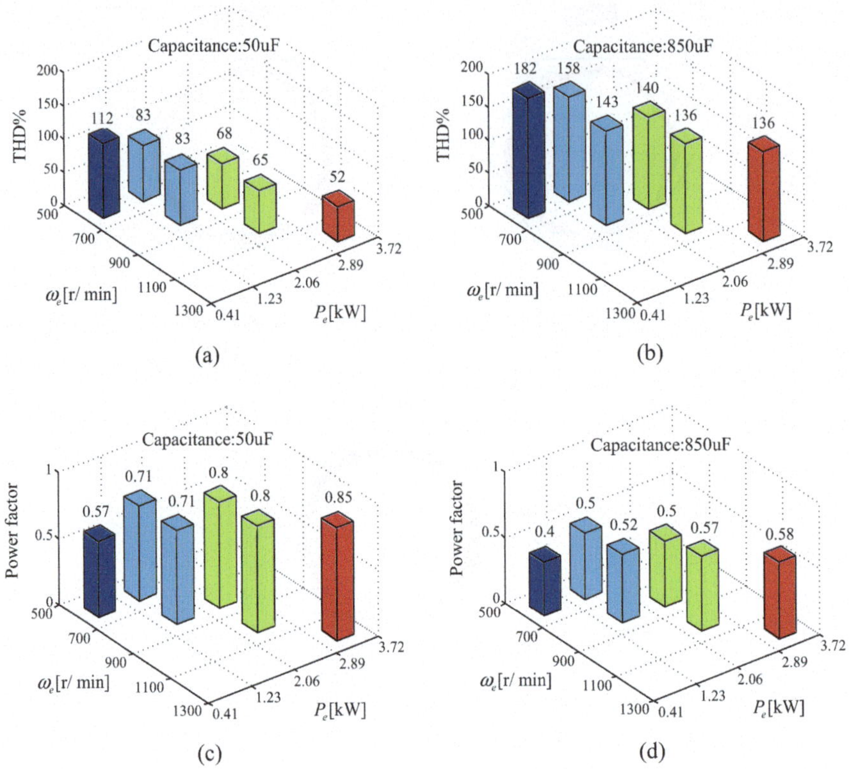

Fig. 6.21 Test results of the grid side power quality in different output power. **a** Phase current THD (inverter with 50 μF capacitor). **b** Phase current THD (inverter with 850 μF capacitor). **c** Power factor (inverter with 50 μF capacitor). **d** Power factor (inverter with 850 μF capacitor)

In order to eliminate the torque ripples, the minimum DC-link voltage is used as the flux-weakening boundary in control mode I. As a result, compared with the flux-weakening method proposed in [11], the fundamental component of the stator currents will increase since larger demagnetization current is needed at the same output condition in flux-weakening region. As shown in Fig. 6.22a, when the speed is 940 r/min, the amplitude of the stator current is 5.65A. Under the same condition, the stator current is 5.25A while the flux-weakening method proposed in [11] is used, as shown in Fig. 6.22b.

The proposed flux-weakening method may decrease the efficiency slightly in control mode I because of the higher amplitude of the stator currents fundamental component. The efficiency of the IPMSM is compared as shown in Fig. 6.23 tested by WT1800. During this operation region, the proposed flux-weakening scheme works in control mode I. The efficiency of the proposed method decreases about 0.6% compared with the method in [11] shown as the two fitting curves in Fig. 6.23. By the way, as the increase of the motor rated power, the decrease of the efficiency will be smaller due to the reduced value of the stator resistance.

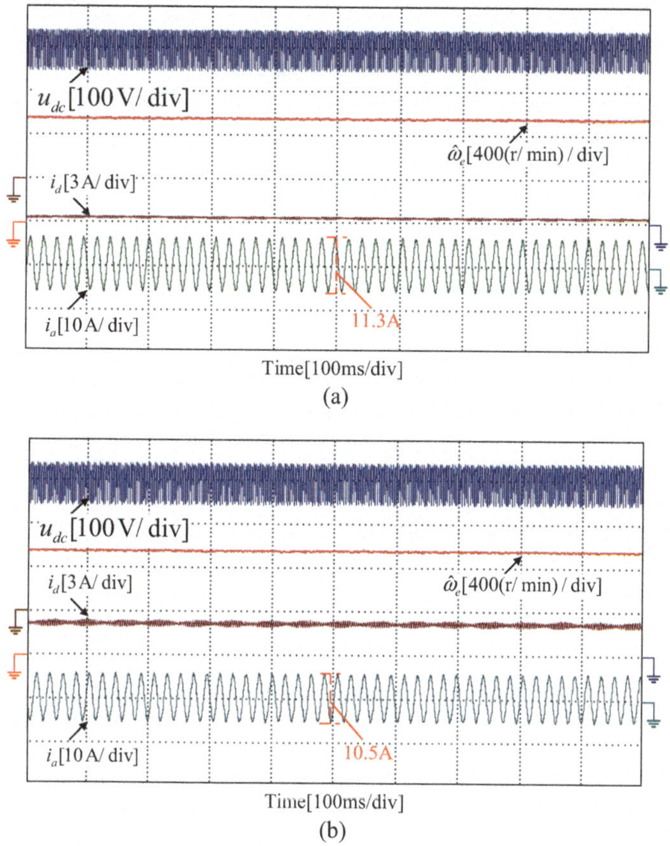

Fig. 6.22 Comparison of the fundamental component of the stator currents at the same operation condition in flux-weakening region. **a** Proposed flux-weakening method in control mode I. **b** Flux-weakening method proposed in [11]

6.6 Summary

This chapter focuses on the flux-weakening control issues existing in the motor drive with reduced DC-link capacitance. A quantitative analysis of the torque ripples in flux-weakening region is given, which will act as the guidance to decide whether to make the full utilization of the DC-link voltage. The adjustable maximum voltage based VCFS is introduced for the reduced DC-link capacitance IPMSM drive. In the minimum DC-link voltage control mode (Mode I), the torque ripples can be significantly reduced at the cost of decreasing the efficiency slightly. In the extended DC-link voltage control mode (Mode II), the DC-link voltage can be further extended, until nearly maximum DC-link voltage utilization. Moreover, the control method only depends on the DC-link voltage and the stator current reference. Although the flux-weakening controller has two control modes, the control system can switch smoothly. Finally, the experimental results have verified the realizability and effectiveness of the control method.

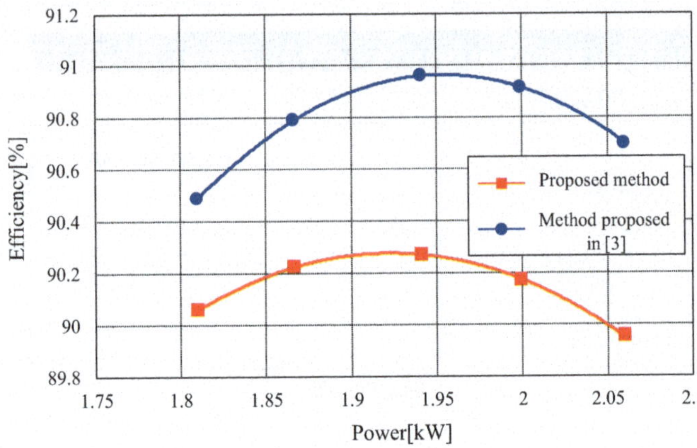

Fig. 6.23 Comparison of the efficiency in flux-weakening region using control mode I

References

1. K. Sang-Hoon, S. Seung-Ki, Voltage control strategy for maximum torque operation of an induction machine in the field-weakening region. IEEE Trans. Ind. Electron. **44**(4), 512–518 (1997)
2. S. Morimoto, Y. Takeda, T. Hirasa, K. Taniguchi, Expansion of operating limits for permanent magnet motor by current vector control considering inverter capacity. IEEE Trans. Ind. Appl. **26**(5), 866–871 (1990)
3. T.M. Jahns, Flux-weakening regime operation of an interior permanent-magnet synchronous motor drive. IEEE Trans. Ind. Appl. **IA-23**(4), 681–689 (1987)
4. L. Harnefors, K. Pietilainen, L. Gertmar, Torque-maximizing field-weakening control: design, analysis, and parameter selection. IEEE Trans. Ind. Electron. **48**(1), 161–168 (2001)
5. S. Kim, J. Seok, Maximum voltage utilization of IPMSMs using modulating voltage scalability for automotive applications. IEEE Trans. Power Electron. **28**(12), 5639–5646 (2013)
6. Z. Dong, Y. Yu, W. Li, B. Wang, D. Xu, Flux-weakening control for induction motor in voltage extension region: torque analysis and dynamic performance improvement. IEEE Trans. Ind. Electron. **65**(5), 3740–3751 (2018)
7. X. Xu, D.W. Novotny, Selection of the flux reference for induction machine drives in the field weakening region. IEEE Trans. Ind. Appl. **28**(6), 1353–1358 (1992)
8. D. Ding, G. Wang, N. Zhao, G. Zhang, D. Xu, Enhanced flux-weakening control method for reduced DC-link capacitance IPMSM drives. IEEE Trans. Power Electron. **34**(8), 7788–7799 (2019)
9. S. Bolognani, M. Zigliotto, Novel digital continuous control of SVM inverters in the overmodulation range. IEEE Trans. Ind. Appl. **33**(2), 525–530 (1997)
10. C. Mademlis, I. Kioskeridis, N. Margaris, Optimal efficiency control strategy for interior permanent-magnet synchronous motor drives. IEEE Trans. Energy Convers. **19**(4), 715–723 (2004)
11. A. Yoo, S. Sul, H. Kim, K. Kim, Flux-weakening strategy of an induction machine driven by an electrolytic-capacitor-less inverter. IEEE Trans. Ind. Appl. **47**(3), 1328–1336 (2011)
12. S. Kim, Y. Yoon, S. Sul, K. Ide, Maximum torque per ampere (MTPA) control of an IPM machine based on signal injection considering inductance saturation. IEEE Trans. Power Electron. **28**(1), 488–497 (2013)

Chapter 7
Motor Loss Based Anti-Overvoltage Control

The DC-link may suffer from overvoltage during motor transient process due to the use of slim film capacitors in the diode rectifier front end. Recently, some researches have been carried out against the overvoltage phenomenon during transient process for motor drives equipped with large electrolytic capacitors [1–7]. One frequently used solution is to add extra hardware, such as switching device controlled braking resistor. Although, this solution is reliable and easy to realize, it will increase the system cost and decrease the power density. Therefore, the anti-overvoltage control scheme realized by software is preferred [8, 9].

7.1 Braking Performance Analysis Under Reduced DC-Link Capacitance

7.1.1 Electrical Power Analysis Under Breaking Process

The PMSM electrical power of the three-phase machine can be expressed as

$$P_{e1} = P_{mech} + P_{Cu} + P_{Fe} + P_{str} + P_m \tag{7.1}$$

where P_{e1} is the electrical power.

Generally, P_{Cu} is the dominated component in all losses, especially when the motor operates below the speed of the flux-weakening [10]. Therefore, only P_{Cu} is considered in this study. The mechanical power and the copper loss can be denoted as

$$P_{mech} = T_e \omega_r \tag{7.2}$$

© The Author(s), under exclusive license to Springer Nature Singapore Pte Ltd. 2021

G. Wang et al., *Reduced DC-link Capacitance AC Motor Drives*,
https://doi.org/10.1007/978-981-15-8566-1_7

$$P_{Cu} = \frac{3}{2} R_s |\mathbf{i_s}|^2. \tag{7.3}$$

In fact, the energy will flow back to the DC-link if $P_{e1} < 0$, here define energy flowing into the motor as the positive direction, and (7.1) could be presented as

$$P_{e1} = T_e \omega_r + \frac{3}{2} R_s |\mathbf{i_s}|^2. \tag{7.4}$$

In maximum torque per ampere (MTPA) control mode, the electromagnetic torque can be denoted as

$$T_e = \frac{3}{2} P_n [\psi_f |\mathbf{i_s}| \sin\theta_M + (L_d - L_q)|\mathbf{i_s}|^2 \sin\theta_M \cos\theta_M]. \tag{7.5}$$

where θ_M is the current angle of the MTPA point.

Then, the magnitude of the stator current can be derived as

$$|\mathbf{i_s}| = \frac{-\frac{3}{2} P_n \psi_f \sin\theta_M - \sqrt{\left(\frac{3}{2} P_n \psi_f \sin\theta_M\right)^2 + 6 T_e P_n \left(L_d - L_q\right) \sin\theta_M \cos\theta_M}}{3 P_n \left(L_d - L_q\right) \sin\theta_M \cos\theta_M} \tag{7.6}$$

Essentially, the direction of energy flow is determined by T_e, and the existence of T_L will only mitigate the phenomenon of overvoltage. Hence, the relationship between the braking torque and the DC-link capacitance is analyzed in the extreme condition of no load.

Combining (7.4) and (7.6), the electrical power in the braking process can be expressed as

$$\begin{aligned} P_{e1} = & \frac{T_e}{P_n} \left(\omega_{ri0} + \frac{T_e P_n}{J} t \right) + \\ & \frac{3}{2} R_s \left(\frac{-\frac{3}{2} P_n \psi_f \sin\theta_M - \sqrt{\left(\frac{3}{2} P_n \psi_f \sin\theta_M\right)^2 + 6 T_e P_n (L_d - L_q) \sin\theta_M \cos\theta_M}}{3 P_n (L_d - L_q) sin\theta_M \cos\theta_M} \right)^2. \end{aligned} \tag{7.7}$$

where ω_{ri0} is the initial rotor speed.

Then, (7.7) can be denoted as

$$P_{e1} = f_1(t, T_e, \theta_M). \tag{7.8}$$

7.1.2 DC-Link Voltage Analysis Under Breaking Process

The rectifier diode will turn off when the energy flows back, then the increase of the DC-link voltage can be expressed as

$$\Delta u_{dcinc} = -\frac{1}{C_{dc}} \int \frac{P_{e1}}{u_{dc}} dt \tag{7.9}$$

where Δu_{dcinc} is the increase of the DC-link voltage.

Further, (7.9) can be denoted as

$$\Delta u_{dcinc} = -\frac{1}{C_{dc}} \int \frac{f_1(t, T_e, \theta_M)}{u_{dci} + \Delta u_{dcinc}} dt \tag{7.10}$$

where u_{dci} is the initial DC-link voltage.

The capacitance can be derived as

$$C_{dc} = \frac{2 f_2(t, T_e, \theta_M)}{2u_{dci} \Delta u_{dcinc} + \Delta u_{dcinc}^2} \tag{7.11}$$

where

$$f_2(t, T_e, \theta_M) = \frac{T_e}{P_n} \omega_{r0} t + \frac{T_e^2}{2J} t^2 + \frac{3}{2} R_s \left(\frac{-\frac{3}{2} P_n \psi_f \sin\theta_M - \sqrt{\left(\frac{3}{2} P_n \psi_f \sin\theta_M\right)^2 + 6 T_e P_n (L_d - L_q) \sin\theta_M \cos\theta_M}}{3 P_n (L_d - L_q) \sin\theta_M \cos\theta_M} \right)^2 t \tag{7.12}$$

Considering the experimental platform and the IPMSM parameters listed in Table 6.1, the quantitative relationship between the braking torque and the DC-link capacitance when the DC-link voltage increasing 150 V is shown in Fig. 7.1. As a result, the following conclusions can be obtained:

(1) Larger DC-link capacitance is needed to maintain secure DC-link voltage when the absolute value of braking torque or deceleration time increases.
(2) By comparing DC-link capacitance at different initial speed (80, 90, and 100% of the rated speed), it can be derived that the motor decelerating from higher initial speed requires larger DC-link capacitance.

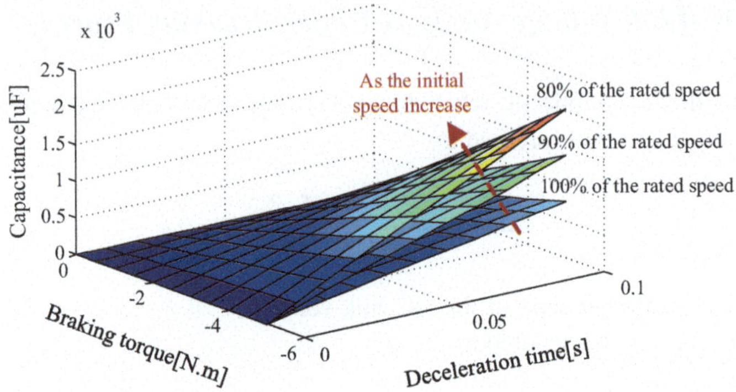

Fig. 7.1 Relationship between the DC-link capacitance, the braking torque and the deceleration time while the DC-link voltage increases

7.2 Motor Loss Based Braking Method

The energy flow during regenerative braking is shown in Fig. 7.2. E_s is the shaft energy, E_f is the energy consumed by friction, E_e is the electromagnetic energy, E_c is the energy consumed by copper loss, and E_{e1} is the electric energy. The core loss can be neglected at normal operating condition. The expressions of the aforementioned energy are denoted as follows,

$$E_s = \frac{1}{2} J \left(\omega_{r\_i}^2 - \omega_{r\_u}^2 \right) \tag{7.13}$$

$$E_f = \int\limits_{\Delta t} F \omega_r^2 dt \tag{7.14}$$

$$E_e = \frac{3}{2} \int\limits_{\Delta t} P_n \left[\psi_f + \left(L_d - L_q \right) i_d \right] i_q \omega_e dt \tag{7.15}$$

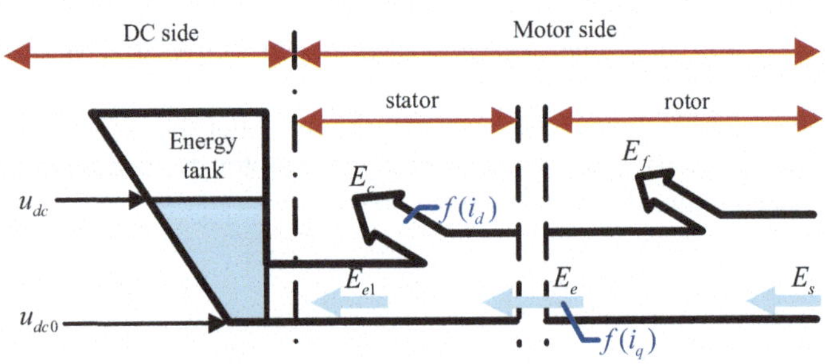

Fig. 7.2 Energy flow during regenerative braking

$$E_c = \int_{\Delta t} \left(i_d^2 + i_q^2\right) R_s dt \tag{7.16}$$

$$E_{e1} = E_e - E_c \tag{7.17}$$

where F is the motor friction coefficient, ω_{r_i} and ω_{r_u} are the initial mechanical angular speed and the ultimate mechanical angular speed during Δt, respectively.

As can be seen in Fig. 7.2, if E_{e1} flows into the DC-link capacitor, the DC-link voltage will increase. Therefore, the only way to regulate the DC-link voltage is to dissipate the regenerative power in the motor, which is the motor loss concept.

Recently, there are some effective methods proposed for active braking based on the concept of motor loss manipulation [11–15]. The characteristics of the above methods are listed in Table 7.1. It should be noted that most of the active braking methods need motor parameter information in order to achieve good dynamic performance during regenerative process. According to the requirements of industrial applications, the active braking method for IPMSM drives without motor parameters is expected for improving the robustness against parameter perturbation. In this chapter, an active braking method based on a defined stator current vector coordinate is introduced for IPMSM drives, and it is different from the existing methods and independent of motor parameters.

Table 7.1 Characteristics of active breaking schemes proposed in recent years

Items Methods	Motor type	Parameters needed	System losses utilized	Way of realization
Method in [11]	IM	Stator resistance Rotor resistance Stator inductance Magnetic inductance	Copper loss Inverter loss	Manipulating losses in each switching period based on loss model
Method in [12]	Six-phase IM	None	Copper loss	Using additional degree of freedom
Method in [13, 14]	SPMSM	Stator resistance Inductance Flux-linkage	Copper loss	Calculating switch speeds using voltage limit, current limit, and power limit
Method in [15]	Switched reluctance motor (SRM)	Stator resistance On-resistance of power device	Copper loss Core loss Inverter loss	Voltage closed-loop control and excitation time calculation

7.3 Stator Current Vector Orientation Based Anti-Overvoltage Control

7.3.1 Principle Analysis

In this section, a novel active braking method is proposed for IPMSM without using any motor parameters.

As shown in Fig. 7.3, the proposed anti-overvoltage control strategy is integrated with vector control scheme. The rotational position θ_M and θ_B are determined by the current reference i_{sref} as can be seen in the dotted box. The current reference i_{sref} is the output of the speed regulator. When the drive system is operating in motoring mode, i_{sref} is greater than zero, and θ_M will be chosen. Otherwise, i_{sref} is lower than zero, and θ_B will be chosen. When the drive system is operating in motoring mode. The stator current will locate at the MTPA curve. Otherwise, the stator current will be allocated by θ_B, which is the current angle in braking process. In this case, θ_B is generated by the voltage controller to guarantee the smoothness of the motor operation in braking process. The d-axis component of stator voltage in the stator current vector coordinate system u_d^{i*} is controlled to zero ($u_{dref}^{i} = 0$) to prevent DC-link overvoltage. The stator current angle in $\alpha\beta$-axes θ_i is obtained by the normalized PLL for the calculation of u_d^{i*}. The amplitude of the stator current is limited to i_{smax}, which is determined by the motor drive system.

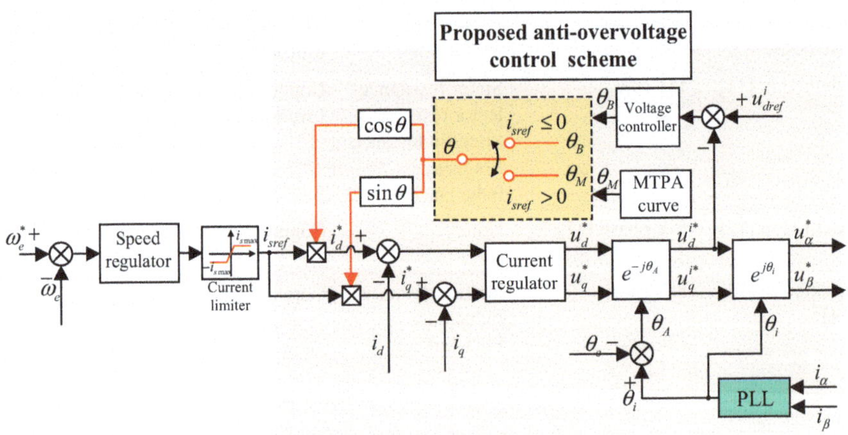

Fig. 7.3 Block diagram of the anti-overvoltage control scheme

7.3.2 Current Trajectory Planning in Braking Process

For electrolytic capacitorless motor drives, the electromagnetic torque should meet the following requirement during regenerative braking process to prevent DC-link overvoltage considering (7.18),

$$-\frac{3P_n R_s|\mathbf{i_s}|^2}{2\omega_e} \leq T_e < 0. \tag{7.18}$$

For the purpose of acquiring satisfactory braking performance and preventing the DC-link overvoltage at the same time, the electromagnetic torque should be kept as

$$T_e = -\frac{3P_n R_s|\mathbf{i_s}|^2}{2\omega_e}. \tag{7.19}$$

The stator current vector can be allocated in the dq-axes considering the expression of electromagnetic torque and (7.19) can be re-expressed as

$$\frac{3}{2}P_n\begin{bmatrix} i_{sref}\sin(\theta_B)\psi_f + \\ (L_d - L_q)i_{sref}^2\sin(\theta_B)cos(\theta_B) \end{bmatrix} = -\frac{3P_n R_s i_{sref}^2}{2\omega_e} \tag{7.20}$$

where i_{sref} is the amplitude of the stator current vector.

Considering (7.20), θ_B is determined by i_{sref} and ω_e as shown in Fig. 7.4. During active braking, θ_B is adjusted automatically according to i_{sref} and ω_e to prevent the DC-link overvoltage and guarantee the smoothness of the motor operation.

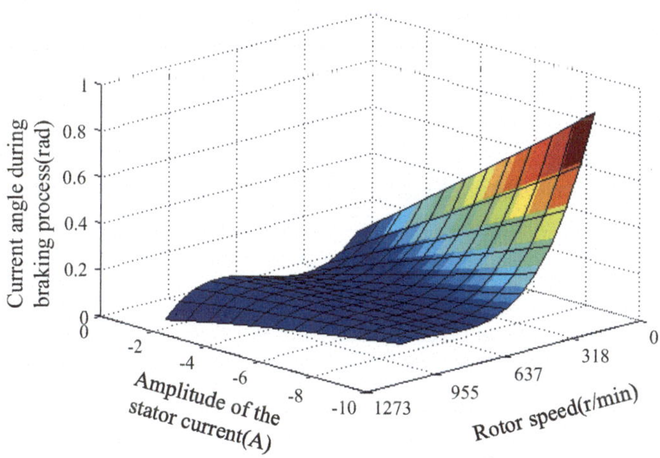

Fig. 7.4 Relationship between the current angle during braking process, the amplitude of the stator current, and the rotor speed

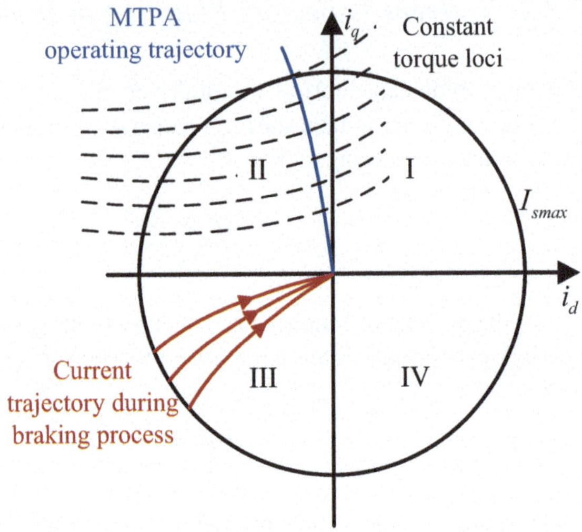

Fig. 7.5 Stator current trajectory for active braking

In regenerative braking mode, the stator current vector will locate at the expected stator current trajectory for active braking shown as the red solid line in quadrant III shown in Fig. 7.5. The stator current trajectory has different starting point determined by i_{sref} and ω_e as analyzed in Fig. 7.4.

Essentially, the energy regenerated by IPMSM is dissipated in the stator resistance. By manipulating the dq-axes currents according to the stator current trajectory, satisfactory braking performance can be obtained while the DC-link overvoltage can be prevented. However, θ_B is related to motor parameters including R_s, L_d and L_q according to (7.20). Therefore, accurate calculation of θ_B from (7.20) is difficult to be achieved since motor parameters will change with the temperature and the magnetic situation. As a result, it is necessary to obtain θ_B in an effective way for practical applications.

7.3.3 Anti-Overvoltage Realization Using Stator Current Vector Orientation

In three-phase diode rectifier IPMSM drives equipped with film capacitors, the average DC-link current can be calculated from the motor side.

$$i_{inv} = \frac{3}{2}\frac{\mathbf{u_s} \cdot \mathbf{i_s}}{u_{dc}} \tag{7.21}$$

Here, a new coordinate frame is denoted as dq^i-axes, in which the stator current vector is located at d^i-axis. For the purpose of preventing the DC-link voltage increase, the average DC-link current cannot be a negative value. Namely, the horizontal component of stator voltage in the defined coordinate frame should not be smaller than zero. As shown in Fig. 7.6, θ_i is the stator current angle in $\alpha\beta$-axes, $\theta_A = \theta_i - \theta_e$ is the angle of stator current vector in dq-axes. The stator voltage vector in the stator current vector coordinate frame can be expressed as

$$\begin{cases} u_d^{i*} = u_d^* \cos\theta_A + u_q^* \sin\theta_A \\ u_q^{i*} = -u_d^* \sin\theta_A + u_q^* \cos\theta_A \end{cases} \tag{7.22}$$

where u_d^* and u_q^* are stator voltage components in dq-axes. u_d^{i*} and u_q^{i*} are stator voltage components in stator current vector coordinate system. The stator current angle in $\alpha\beta$-axes obtained by PLL is shown in Fig. 7.7.

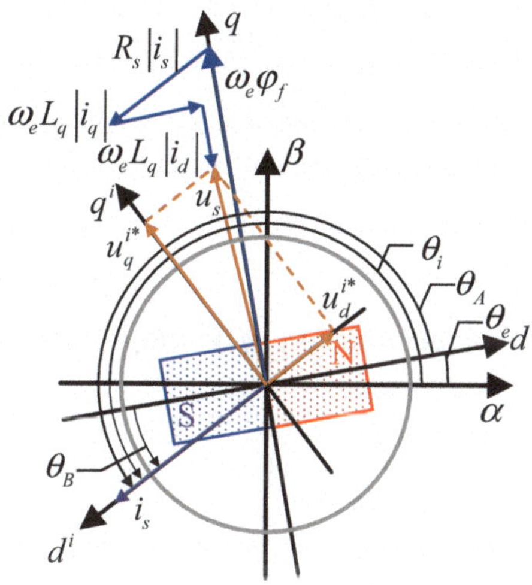

Fig. 7.6 Stator voltage components in stator current vector coordinate frame

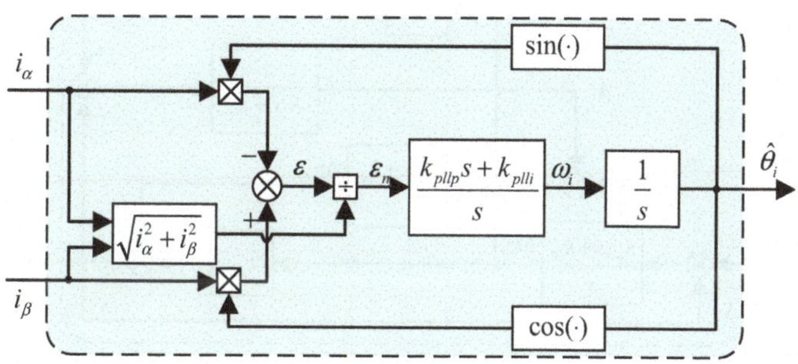

Fig. 7.7 Block diagram of the PLL for acquiring θ_i

Compared with the traditional structure, the stator current is normalized to eliminate its effect on the bandwidth of the PLL control loop. Consequently, the transfer function of the PLL after linearization can be derived as

$$\frac{\hat{\theta}_i}{\theta_i} = \frac{k_{pllp}s + k_{plli}}{s^2 + k_{pllp}s + k_{plli}} \tag{7.23}$$

The parameters of the PLL can be designed by adopting poles assignment. Generally, the poles can be set on the negative real axis. In this study, choosing $k_{pllp} = 2\rho$, $k_{plli} = \rho^2$ and the bandwidth of the PLL should be higher than that of the current loop. When ρ is set as 600, the bandwidth of the PLL is 1000 Hz.

Ultimately, the current angle in braking process can be derived by controlling the stator voltage vector in the stator current vector coordinate frame

$$\theta_B = \frac{k_{p\theta}s + k_{i\theta}}{s}\left(u^i_{dref} - u^{i*}_d\right) \tag{7.24}$$

where $k_{p\theta}$ and $k_{i\theta}$ are the coefficients of the anti-overvoltage PI controller, u^i_{dref} is the reference of the horizontal component of stator voltage in stator current vector coordinate system.

7.3.4 Parameters Determination of Voltage Controller

Choosing one operating point as $\theta_{B0} = 0$, $u^*_{d0} = 0$, $u^*_{q0} = |u_s|$, $u^{i*}_{d0} = 0$, $u^{i*}_{q0} = |u_s|$. Under this circumstance, the small signal model of the proposed anti-overvoltage control scheme is shown in Fig. 7.8.

Therefore, the transfer function of the anti-overvoltage control scheme can be derived

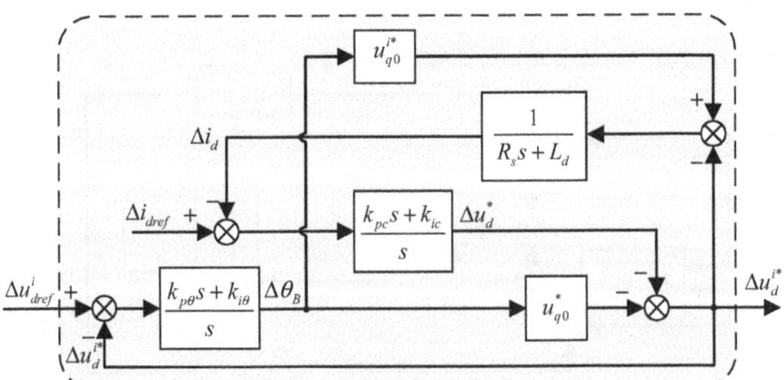

Fig. 7.8 Small signal Block diagram of the anti-overvoltage control scheme

$$\frac{\Delta u_d^{i*}}{\Delta u_{dref}^{i*}} = \frac{-k_{p\theta}|\mathbf{u_s}|s^2 + (-k_{i\theta}|\mathbf{u_s}| + |\mathbf{u_s}|\omega_{cb}k_{p\theta})s + k_{i\theta}|\mathbf{u_s}|\omega_{cb}}{(1 - k_{p\theta}|\mathbf{u_s}|)s^2 + (k_{p\theta}|\mathbf{u_s}|\omega_{cb} - k_{i\theta}|\mathbf{u_s}| + \omega_{cb})s + k_{i\theta}\omega_{cb}|\mathbf{u_s}|} \tag{7.25}$$

where Δu_d^{i*} and Δu_{dref}^{i*} are the disturbance components of u_d^{i*} and u_{dref}^{i*}, respectively. ω_{cb} is the bandwidth of the current loop, while the coefficients of the current regulator (k_{pc}, k_{ic}) is designed to eliminate the electromagnetic time of the current loop.

In order to guarantee the system stability, Routh-Hurwitz stability criterion is considered

$$\begin{cases} 1 - k_{p\theta}|\mathbf{u_s}| > 0 \\ k_{p\theta}|\mathbf{u_s}|\omega_{cb} - k_{i\theta}|\mathbf{u_s}| + \omega_{cb} > 0 \\ k_{i\theta}|\mathbf{u_s}|\omega_{cb} > 0 \end{cases} \tag{7.26}$$

The coefficients of the PI controller in anti-overvoltage scheme should meet the following conditions

$$\begin{cases} 0 < k_{p\theta} < \frac{1}{|\mathbf{u_s}|} \\ 0 < k_{i\theta} < \frac{\omega_{cb}}{|\mathbf{u_s}|} \end{cases}. \tag{7.27}$$

The maximum amplitude of the stator voltage vector is considered to guarantee the system stability in whole speed region. Here, the bandwidth of the current loop is set as 200 Hz, which can be set as several hundred Hz as well. As a result, the pole map of the control system can be obtained as Fig. 7.9, when the coefficients of the PI controller are set according to (7.27). As can be seen, the increase of $k_{p\theta}$ will reduce the settling time of the controller, while the system damping reduces with the increase of $k_{i\theta}$. Therefore, choosing $k_{p\theta} = 0.0023$, $k_{i\theta} = 4$.

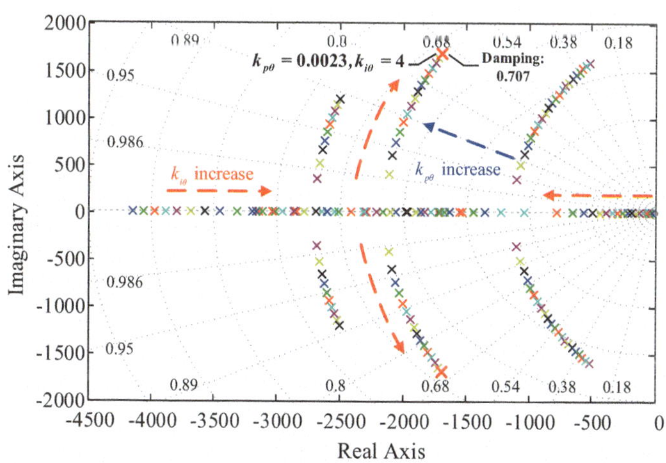

Fig. 7.9 Pole map of the anti-overvoltage control scheme

In order to promise better speed transient performance, the speed regulator is designed using the method proposed in [16, 17]. The open-loop transfer function of the speed loop can be denoted as

$$H(s) = \left(k_{ps} + \frac{k_{is}}{s}\right)\frac{1}{T_{cc}s + 1}\frac{1}{Js} \tag{7.28}$$

where T_{cc} is the time constant of the current loop, k_{ps} and k_{is} are the coefficients of the speed regulator.

With the definition of the cutoff frequency ω_{sc} and phase margin φ_m of the speed loop, the coefficients of speed regulator can be derived.

$$k_{ps} = \sqrt{\left(1 + T_{cc}^2\omega_{sc}^2\right)\left(J^2\omega_{sc}^2\right)}\sin\left(\varphi_m - \arctan\left(\frac{1}{\omega_{sc}T_{cc}}\right) + \frac{\pi}{2}\right), \tag{7.29}$$

$$k_{is} = \omega_{sc}\sqrt{\left(1 + T_{cc}^2\omega_{sc}^2\right)\left(J^2\omega_{sc}^2\right)}\cos\left(\varphi_m - \arctan\left(\frac{1}{\omega_{sc}T_{cc}}\right) + \frac{\pi}{2}\right). \tag{7.30}$$

In order to simplify the calculation process, the following assumption is made that $T_{cc}\omega_{sc} << 1$. Finally, the coefficients of the speed regulator can be derived as follows,

$$\begin{cases} k_{ps} = J\omega_{sc}\sin\varphi_m \\ k_{is} = J\omega_{sc}^2\cos\varphi_m \end{cases}. \tag{7.31}$$

7.3.5 Experimental Results

The anti-overvoltage control scheme is verified on a 2.2-kW IPMSM, supplied by a commercial three-phase diode rectifier drive with film capacitors shown in Fig. 6.13. The targeted application of the anti-overvoltage scheme is heating, ventilation and air conditioning (HVAC) system. Parameters of the test motor are listed in Table 6.1. The load motor is working as a generator with its output connecting to a resistance box. Consequently, the load will increase with the motor speed, which is the load type of the HVAC system. The coefficients of the speed regulator are set as $k_{ps} = 70$ and $k_{is} = 12000$, and the coefficients of the current regulator are set as $k_{pc} = 101$ and $k_{ic} = 5162$.

The comparison of the DC-link voltage and the stator current during deceleration is shown in Fig. 7.10. Under this condition, the hardware brake unit is used to limit the DC-link voltage within the safety value. In the experiment, the activating voltage

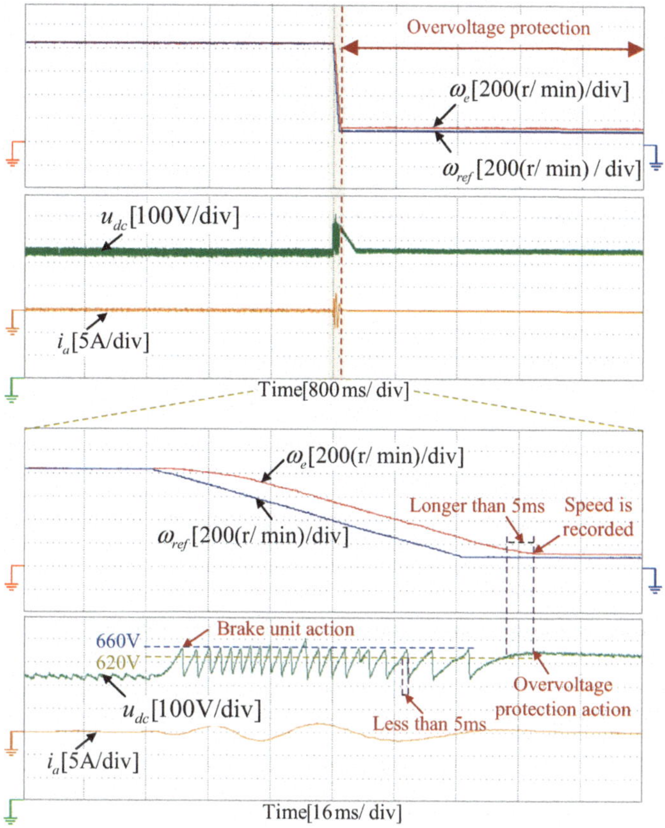

(a)

Fig. 7.10 Experimental comparison of the speed decrease from 900 to 100r/min in no-load condition. **a** Without anti-overvoltage control. **b** With proposed anti-overvoltage control. **c** Without anti-overvoltage control at the same speed change rate. (d) Speed reference slop is set as -55.6r/s^2. **e** Speed reference slop is set as $-16.7\ \text{r/s}^2$

of the brake unit is set at 660 V, and the threshold of the overvoltage protection is set at 620 V. The overvoltage protection execution cycle is 1 ms, and the protection will be activated to stop the Interrupt Service Routine of the vector control when the overvoltage protection flag is maintained for five execution cycles (5 ms).

At the beginning, the motor is operating at MTPA point, and the DC-link overvoltage emerges as the speed reference changes from 900 to 100 r/min within 80 ms as shown in Fig. 7.10a. The brake unit is turned on to consume the regenerative energy once the DC-link voltage reaches 660 V. Due to the adoption of the slim film capacitor, the change of the DC-link voltage is fast. Thus, the overvoltage protection will not be activated when the electrical level of the protection flag is maintained less than 5 ms. When the motor speed decrease process is nearly finished, the

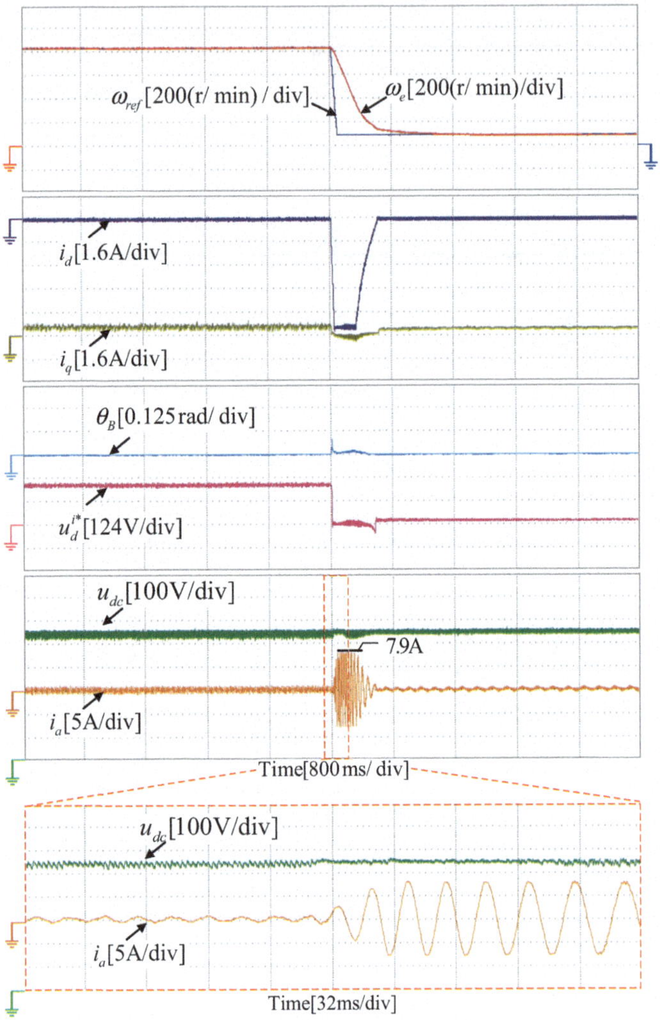

(b)

Fig. 7.10 (continued)

overvoltage protection will be activated since the protection flag is maintained longer than 5 ms. At this time, the speed is recorded instantly and the motor enters into the freewheeling mode. By modifying the *dq*-axes currents, the overvoltage problem can be solved as shown in Fig. 7.10b. As can be seen, θ_B is calculated immediately during deceleration process, u_d^{i*} is controlled to zero immediately to prevent the energy flow back. In the experiment, i_{smax} is chosen as 7.9 A which is the rated current of the test IPMSM. As can be seen in Fig. 7.10b, the peak value of the a-phase current can reach 7.9 A during speed decreasing process. Therefore,

Fig. 7.10 (continued)

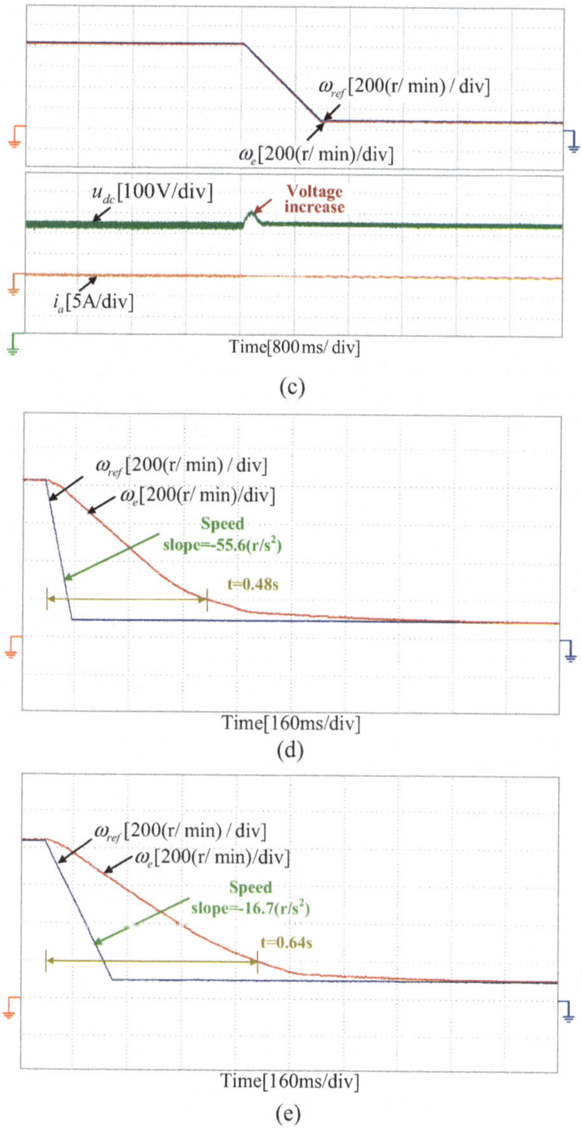

the proposed anti-overvoltage scheme can maximize the copper loss during speed decelerating process. The speed adjustment time is 1.2 s when the speed decreases from 900 to 100 r/min. It should be noted that, the large speed reference change rate is set to accelerate the braking process. On the one hand, the speed reference change rate should give enough margin for the increase of the real speed change rate. On the other hand, large difference between the speed reference and the real speed during braking process is beneficial for the increase of the copper loss. Figure 7.10c shows the experimental result when the speed adjustment time is set as the same in Fig. 7.10b (1.2 s) without the proposed anti-overvoltage control scheme. As can be seen, the DC-link voltage increases from 540 to 610 V during the speed deceleration process. Figure 7.10d and e show the experimental results when the deceleration slop of the speed reference is set as -55.6 r/s^2 and -16.7 r/s^2, respectively. The speed reference decelerates from 900 to 100 r/min. As can be seen in Fig. 7.10d, the deceleration time is 0.48 s when the speed reaches 200 r/min. If the speed slop is reduced to -16.7 r/s^2, the deceleration time is increased to 0.64 s as shown in Fig. 7.10e. It should be noted that, this deceleration experiments are done with no-load, in which condition the overvoltage phenomenon is more likely to emerge. If the motor decelerates with load, the braking process will be easier and faster.

Figure 7.11 shows the experimental results of the sudden load reduction while the motor speed is 900 r/min. When the load reduces from 80% of the rated value to 0 N·m suddenly, the drive system will operate in regenerative braking mode. Under this condition, the overvoltage fault will happen as shown in Fig. 7.11a. Figure 7.11b and c show the 80 and 100% rated load reduction when the anti-overvoltage control scheme is applied. As can be seen, when the q-axis current is lower than zero, the dq-axes current is reallocated by θ_B. The absolute value of the q-axis current is small mainly to restrict the electromagnetic torque, the absolute value of the d-axis current is increased to consume more energy in stator resistance. The DC-overvoltage during braking process has been prevented effectively and the speed settling time is 400 ms.

When the speed reference changes from 900 r/min (corresponding to 80% rated load) to 100 r/min (corresponding to 1% rated load) within 80 ms, the overvoltage phenomenon will still occur as shown in Fig. 7.12a. Figure 7.12b and c show the deceleration under 80 and 100% initial load with the proposed anti-overvoltage control scheme. The DC-link overvoltage can be avoided, and the speed settling time is 400 ms. Besides, the adjustment process is faster than no-load condition compared with Fig. 7.10b.

7.4 Energy Control Error Analysis of Braking Scheme

Conventionally, the DC-link overvoltage can be prevented by the aforementioned voltage controller effectively. However, it may fail in electrolytic capacitorless motor drives due to the small capacity of the DC-link capacitor and the undesirable energy error.

Due to the time delay of the q-axis current reference and the real q-axis current, the current control error is inherent during transient process, which can be expressed as

$$i_{qerr} = i_q^* - i_q \tag{7.32}$$

where i_{qerr} is the q-axis current control error.

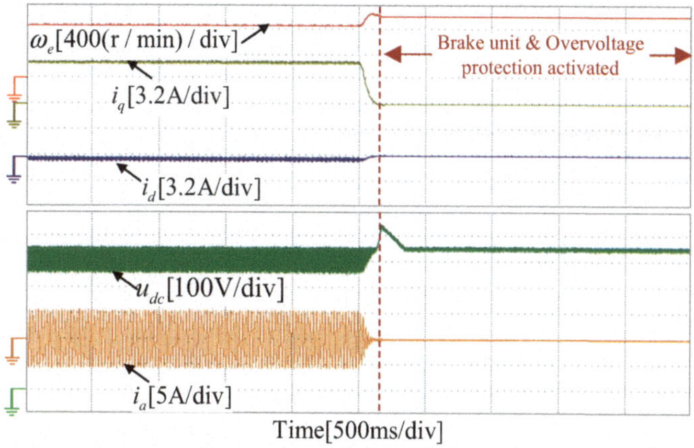

(a)

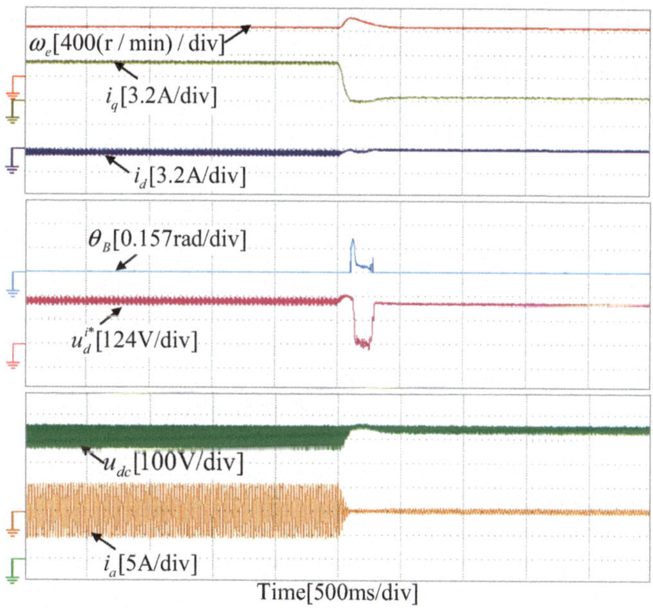

(b)

Fig. 7.11 Experimental comparison of the sudden load reduction. **a** Load reduces from 80% rated value to 0 N·m without anti-overvoltage control. **b** Load reduces from 80% rated value to 0 N·m with proposed anti-overvoltage control. **c** Load reduces from 100% rated value to 0 N·m with proposed anti-overvoltage control

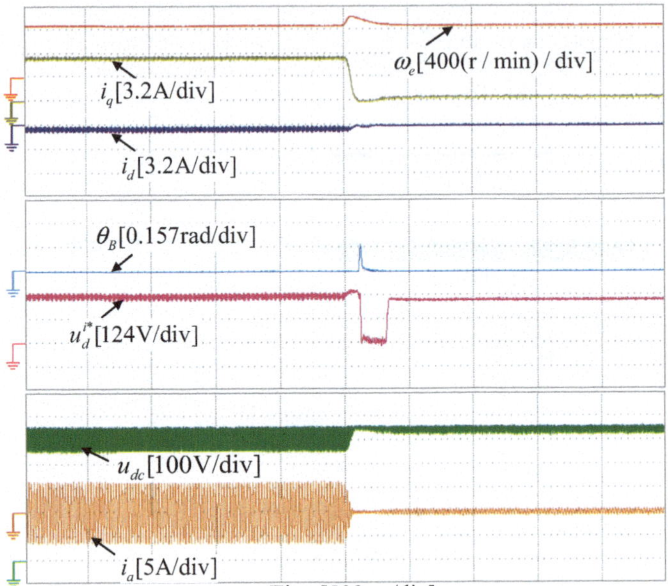

Fig. 7.11 (continued)

The energy error produced by the q-axis current control error from t_1 to t_2 can be derived as

$$E_{err} = \frac{3}{2}\int_{t_1}^{t_2} P_n \psi_f \omega_e i_{qerr} dt \tag{7.33}$$

where E_{err} is the energy error.

Generally, the time scale from t_1 to t_2 is very small compared with the speed deceleration time, and ω_e can be regarded as an instant value denoted as ω_{e0}. Therefore, the energy error can be represented as

$$E_{err} = \frac{3}{2} P_n \psi_f \omega_{e0} \int_{t_1}^{t_2} i_{qerr} dt \tag{7.34}$$

The energy error will affect the DC-link voltage during regenerative process

$$E_{err} = \frac{1}{2} C_{dc}[u_{dc}^2 - u_{dci}^2] \tag{7.35}$$

Ultimately, the DC-link voltage error caused by the q-axis current control error can be expressed as

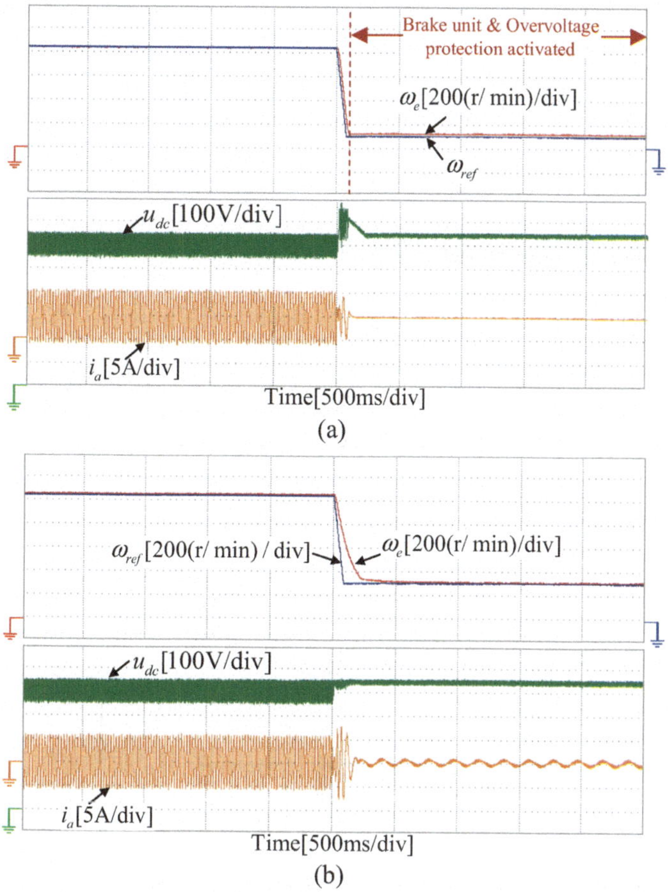

Fig. 7.12 Experimental comparison of the deceleration from 900 to 100r/min under load condition. **a** 80% rated load without anti-overvoltage control. **b** 80% rated load with proposed anti-overvoltage control. **c** 100% rated load with proposed anti-overvoltage control

$$u_{\mathrm{dcerr}} \approx \frac{3P_n\psi_f\omega_{e0}\int_{t_1}^{t_2} i_{qerr}dt}{2C_{dc}u_{dci}} \tag{7.36}$$

where $u_{dcerr} = u_{dc} - u_{dci}$ is the DC-link voltage control error.

The block diagram of the q-axis current control error is illustrated in Fig. 7.13, in which $1/(T_{s\omega}s + 1)$ is the closed-loop transfer function of the current loop. During transient process, i_{qerr} will cause the DC-link voltage control error u_{dcerr}. Consequently, the voltage controller will adjust the q-axis current reference i_q^* for the purpose of stabilizing the DC-link voltage. However, due to the existence of the inherent q-current control error, the voltage controller cannot regulate the DC-link voltage effectively through the q-axis current reference adjustment. When the large-volume electrolytic capacitor is applied, the voltage controller can achieve perfect

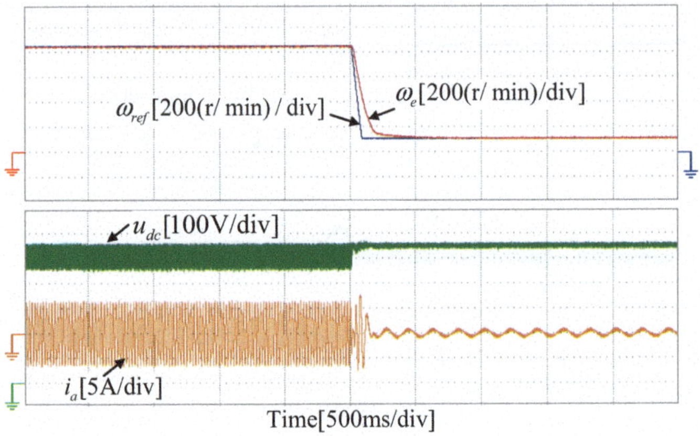

(c)

Fig. 7.12 (continued)

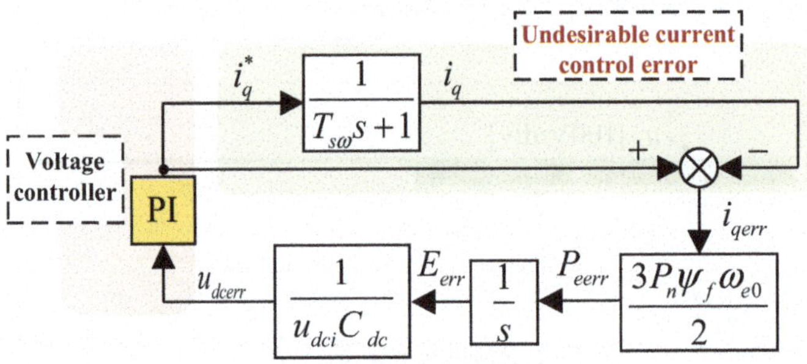

Fig. 7.13 Signal flow graph of the current error regulated by DC-link voltage controller

performance. However, little energy error will cause large u_{dcerr} in motor drives equipped with slim film capacitors. Figure 7.14 shows the DC-link voltage error caused by the q-axis current control error. During one fluctuation period, i_d can be regarded as a constant since it is only used for loss consumption. During braking process, u_{dc} will increase if $i_q^* > i_q$ (from t_1 to t_2) in region 1, otherwise u_{dc} will decrease if $i_q^* < i_q$ in region 2.

Figure 7.15 is the relationship of the DC-link voltage error, the DC-link capacitance and the motor speed considering the experimental electrolytic capacitorless IPMSM drive according to (7.36). The motor parameters are listed in Table 6.1. During 1.6 ms, different q-axis current control error is considered under different speeds and capacitance. As can be seen, when the DC-link capacitance increases, the DC-link voltage error caused by current control error will decrease under the same motor speed. Quantificationally, if a 50 μF DC-link capacitor is used, the voltage

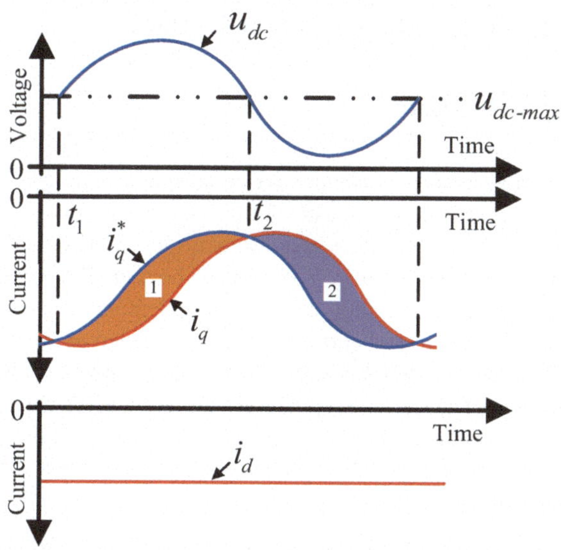

Fig. 7.14 Voltage control error caused by the q-axis current error

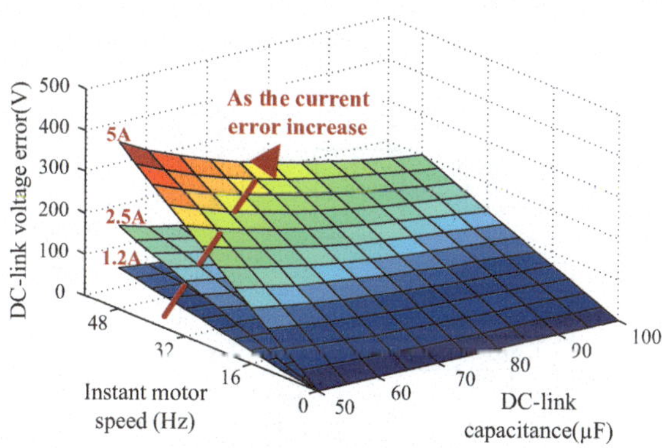

Fig. 7.15 Relationship of the DC-link voltage error, the DC-link capacitance and the instant motor speed

error reaches 200 V when the q-axis current error is 5 A at the operating frequency of 45 Hz. As a result, the voltage controller may fail when using slim film capacitor, since the energy error cannot be controlled effectively, especially when the d-axis current is large.

7.5 Dual Anti-Overvoltage Control Method

7.5.1 Principle Analysis

The proposed anti-overvoltage scheme is based on the system loss concept, and it is enhanced for the applications of electrolytic capacitorless drives. The dual anti-overvoltage control scheme for electrolytic capacitorless IPMSM drives with coefficient autoregulation is shown in Fig. 7.16. As can be seen, the q-axis current limit $i_{q\text{-}max}$ is determined by a PI regulator, and the coefficient autoregulation strategy is integrated. In addition, the d-axis current is used to accelerate the braking process by increasing the copper loss quickly. The setting of the two voltage limits $u_{dc\text{-}l1}$ and $u_{dc\text{-}l2}$ will be analyzed. The speed and position estimation method presented in [18] is used for achieving position sensorless control. To conclude, for the applications of electrolytic capacitorless drives, the improvement of the proposed anti-overvoltage control scheme is listed as follows:

(1) The dual anti-overvoltage scheme is proposed to mitigate the effect of the energy control error on the DC-link voltage control performance.
(2) The coefficients of the DC-link voltage controller are theoretically designed to acquire sufficient stability margin.
(3) The copper loss can be maximized instantly and maintained in braking process.

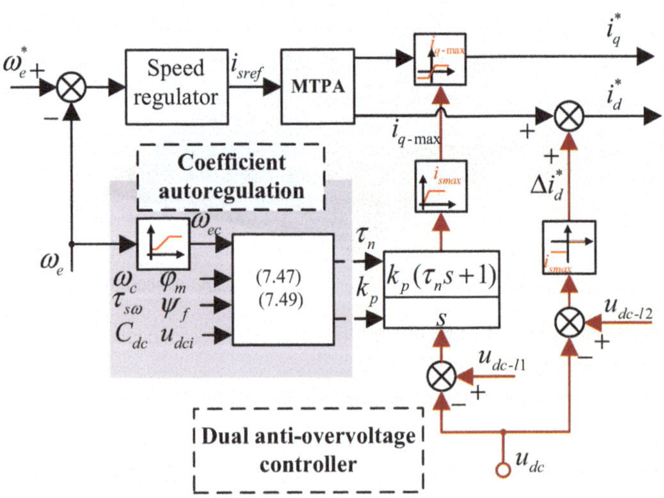

Fig. 7.16 Block diagram of the proposed braking scheme

7.5.2 Realization of Dual Anti-Overvoltage Control Method

The energy flow during regenerative braking is shown in Fig. 7.17. The shade area in the trapezium represents the energy stored in the DC-link capacitor. The height of the shade area in trapezium is the value of DC-link voltage. Switches 1, 2 and 3 are controlled by the *dq*-axis currents, i_d and i_q. E_s is the shaft energy, E_l is the load energy, E_f is the energy consumed by friction, E_c is the energy consumed by copper loss, and E_{e1} is the electric energy. E_{Fe} is the energy consumed by core loss, which is ignored. The smaller E_s will make the control of the DC-link voltage easier, since the motor load can provide extra braking toque for active braking, and this part of torque will not cause the DC-link voltage increase.

During regenerative braking, there are two aims expected. Firstly, the DC-link voltage should be controlled at a secure value. Secondly, the motor braking process should be as fast as possible to meet the requirement of high-dynamic applications. Based on the energy flow during regenerative braking as shown in Fig. 7.17, there are four cases can be derived according to the states of Switch 1, Switch 2 and Switch 3 (directly related to i_d and i_q):

Case 1: $i_d = 0$ & $i_q = 0$. Under this condition, Switch 1, Switch 2 and Switch 3 are all off. The regenerative energy can only be consumed by motor friction (E_f). Although the DC-link voltage does not increase, the braking process is very slow.

Case 2: $i_d \neq 0$ & $i_q = 0$. Under this condition, Switch 1 is off, Switch 2 and Switch 3 are on. The regenerative energy cannot flow into the DC-link capacitor. The energy in the capacitor will be consumed by E_f and E_c. That means the DC-link voltage will decrease quickly, but the motor braking process is still very slow.

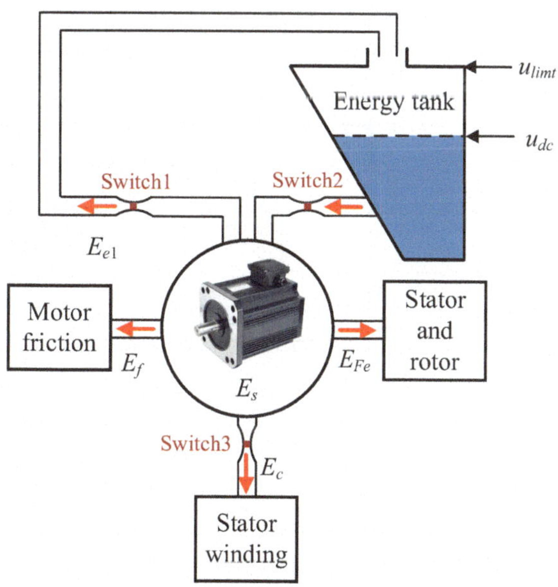

Fig. 7.17 Energy flow during regenerative braking

Case 3: $i_d = 0$ & $i_q < 0$. Under this condition, Switch 1, Switch 2 and Switch 3 are all on. The braking energy flows into the DC-link capacitor and then is consumed by E_f and E_c.

Case 4: $i_d < 0$ & $i_q < 0$. Under this condition, Switch 1, Switch 2 and Switch 3 are all on. Large regenerative braking energy flows into the DC-link capacitor, but is quickly consumed by increased E_c.

As a result, case 4 is the best solution for active braking. The proposed method is designed based on case 4 to prevent the DC-link overvoltage while improving the motor dynamic performance.

By the manipulation of *dq*-axis currents, case 4 can be realized. The DC-link overvoltage can be prevented while guaranteeing the motor dynamic performance. The proposed coefficients autoregulation method is integrated into the controller specially for motor drive with slim film capacitor in order to guarantee the effective manipulation of the *dq*-axis currents.

7.5.3 Analysis of Energy Control Error

The voltage control error of the proposed dual anti-overvoltage control scheme is analyzed in Fig. 7.18. Also, $u_{dc\text{-}l2}$ is higher than any magnitude that u_{dc} can ever arrive during motoring mode. When u_{dc} reaches $u_{dc\text{-}l1}$, the DC-link voltage controller will

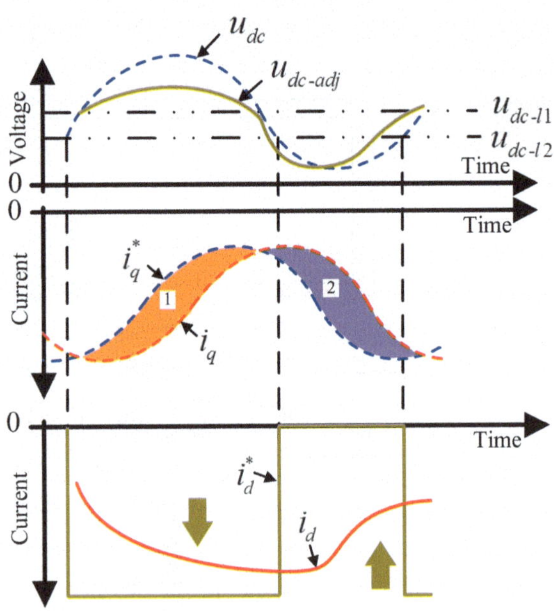

Fig. 7.18 Voltage control error of the dual anti-overvoltage control scheme

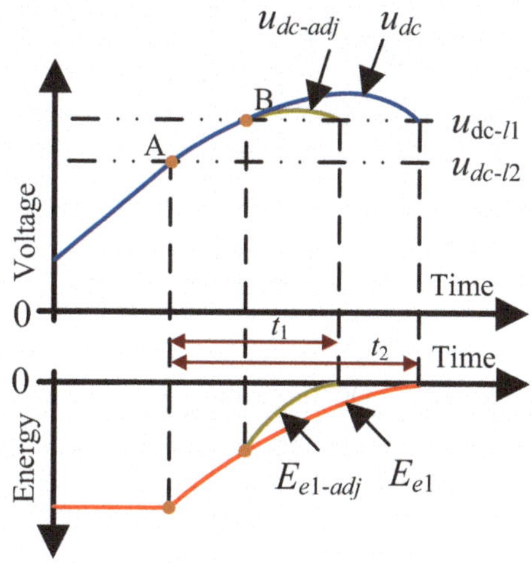

Fig. 7.19 Energy control error of the dual anti-overvoltage control scheme

be activated to limit the q-axis current, and this part is the same as the aforementioned method. The comparison between u_{dc} and $u_{dc\text{-}l2}$ is added to control the d-axis current reference. Once u_{dc} exceeds $u_{dc\text{-}l2}$, the d-axis current reference will change in step type to the negative maximum value. This kind of manipulation has two effects: (1) u_{dc} can be reduced effectively to $u_{dc\text{-}adj}$ in region 1 as can be seen in Fig. 7.18, (2) the d-axis current can be maximized as fast as possible. In region 2, the DC-link voltage decreases with $u_{dc\text{-}l2}$ due to the regulation of the d-axis current. However, this is not important since the maximum value of DC-link voltage suppression is the target for anti-overvoltage control.

In practical applications, the gap between $u_{dc\text{-}l1}$ and $u_{dc\text{-}l2}$ requires a trail-and-error approach considering the energy control error. Figure 7.19 shows the energy control ability of the proposed anti-overvoltage scheme. Once u_{dc} reaches point A, E_{e1} can be reduced as one axis current component is controlled. If u_{dc} increases to point B further, another axis current component will be controlled to reduce E_{e1} simultaneously. As can be seen, compared with the conventional method, the dual control scheme can reduce the energy regulation time from t_2 to t_1. Small gap between $u_{dc\text{-}l1}$ and $u_{dc\text{-}l2}$ will reduce the energy regulation time. However, if the gap is set excessively smaller, the copper loss will decrease.

7.5.4 *Voltage Controller Coefficient Autoregulation*

The relationship between the electric power and the DC-link voltage can be expressed as

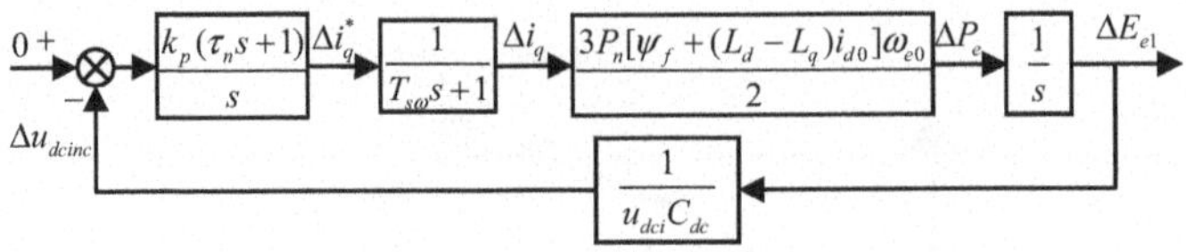

Fig. 7.20 Small signal diagram of the anti-overvoltage control scheme

$$E_{e1} = \frac{1}{2}C_{dc}[u_{dc}^2 - u_{dci}^2] \tag{7.37}$$

As a result, the increase of the DC-link voltage can be denoted as

$$\Delta u_{dcinc} \approx \frac{E_{e1}}{C_{dc}u_{dci}}. \tag{7.38}$$

The small signal block diagram of the anti-overvoltage control system is shown in Fig. 7.20. The open-loop transfer function of the proposed anti-overvoltage scheme can be derived as

$$G(s) = \frac{K(\tau_n s + 1)}{s^2(T_{s\omega}s + 1)} \tag{7.39}$$

$$K = \frac{3P_n\left[\psi_f + (L_d - L_q)i_{d0}\right]\omega_{e0}k_p}{2C_{dc}u_{dc0}} \tag{7.40}$$

where k_p and τ_n are the coefficients of the voltage controller.

The cutoff frequency and the phase margin of the voltage control loop can be derived from (7.40),

$$|G(j\omega_c)| = \frac{K\sqrt{(\tau_n\omega_c)^2 + 1}}{\omega_c^2\sqrt{(T_{s\omega}\omega_c)^2 + 1}} = 1 \tag{7.41}$$

$$\angle G(j\omega_c) = \arctan(\tau_n\omega_c) - \pi - \arctan(\tau_{s\omega}\omega_c) = \varphi_m - \pi \tag{7.42}$$

where ω_c and φ_m are the cutoff frequency and the phase margin, respectively.

By solving (7.41) and (7.42), the coefficients of the voltage controller can be derived as

$$\tau_n = \frac{\tan(\varphi_m) + T_{s\omega}\omega_c}{\omega_c[1 - \tan(\varphi_m)T_{s\omega}\omega_c]} \tag{7.43}$$

$$k_p = \frac{2Cu_{dci}\omega_c^2[1 - tan(\varphi_m)T_{s\omega}\omega_c]\sqrt{(T_{s\omega}\omega_c)^2 + 1}}{3P_n[\psi_f + (L_d - L_q)i_{d0}]\omega_{e0}\sqrt{\begin{array}{l}[tan(\varphi_m) + T_{s\omega}\omega_c]^2\\ +[1 - tan(\varphi_m)T_{s\omega}\omega_c]^2 + 1\end{array}}} \tag{7.44}$$

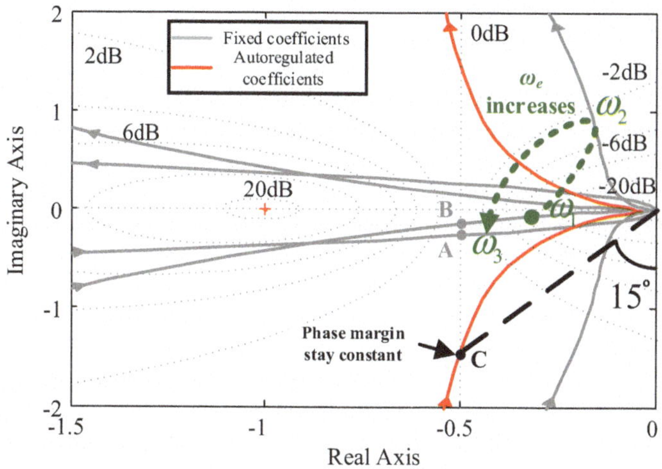

Fig. 7.21 Nyquist plots of the control system using fixed coefficients and autoregulated coefficients

Neglecting the influence of the d-axis current, k_p can be simplified as

$$k_p = \frac{2Cu_{dci}\omega_c^2[1 - \tan(\varphi_m)T_{s\omega}\omega_c]\sqrt{(T_{s\omega}\omega_c)^2 + 1}}{3P_n\psi_f\omega_{e0}\sqrt{[tan(\varphi_m) + T_{s\omega}\omega_c] + [1 - \tan(\varphi_m)T_{s\omega}\omega_c]^2 + 1}} \tag{7.45}$$

As can be seen in (7.45), except for the motor speed ω_{e0}, parameters such as ψ_f and C_{dc} may not change dramatically with the drive operating condition. By considering the motor speed in k_p, the control system can avoid the influence of the motor speed variation. It means that, once the cutoff frequency and the phase margin are set, the coefficients of the voltage controller will be regulated automatically to maintain the phase margin of the system. In practical control system, the motor speed information used in k_p should be limited to a suitable range. Figure 7.21 shows the Nyquist plots of the control system using fixed coefficients and autoregulated coefficients, respectively. If fixed controller coefficients are used, the Nyquist plot is shown as the gray line. The Nyquist plot changes with the motor speed, and the direction is depicted as the green line. When the motor speed increases from ω_1 to ω_2, the system phase margin will increase. However, as the speed further increases from ω_2 to ω_3, the phase margin will decrease. That means, the system phase margin will change with the motor speed if fixed controller coefficients are used. When autoregulated coefficients are used, the Nyquist plot of the control system is shown as the red line. The system frequency characteristic will not change with the motor speed, and the phase margin can be maintained to the designed value. In this control system, the phase margin is designed as 75°.

7.5.5 Experimental Results

The proposed anti-overvoltage control scheme has been verified on a 2.2-kW IPMSM platform, supplied by a three-phase diode rectifier drive with film capacitors shown in Fig. 6.13. The inverter includes a 50 μF film capacitor, power board and control board. The AC input of the drive is 380-Vrms (50 Hz). Parameters of the test motor are listed in Table 6.1 and the amplitude of the stator current vector is limited to 7.9 A. The load motor is used as a generator and consumes energy at load resistors. All algorithms are implemented in a low-cost STM32F103 ARM with 72 MHz clock frequency. The inverter switching frequency is set as 6 kHz, the same as the current sampling frequency.

Figure 7.22 shows the control performance of the system loss based anti-overvoltage control scheme [4] applied in electrolytic capacitorless IPMSM drives. As can be seen in Fig. 7.22a, the DC-link voltage is controlled around 640 V ($u_{\text{dc-max}}$) during the speed decrease process and the deceleration time is 0.6 s from 45 to 5 Hz. However, the magnitude of the DC-link voltage fluctuation trends to increase along with the increase of the d-axis current. The reason is that, the q-axis current will be increased as the increase of the copper loss. The q-axis current control error will also be enlarged without regulator coefficient autoregulation. When the copper loss is further increased for accelerating the deceleration process in Fig. 7.22b, the DC-link voltage will further increase when the d-axis current reaches the negative maximum value. The enlarged view of the DC-link voltage fluctuation is shown in Fig. 7.22c. As can be seen, when i_q^* is greater than i_q in region 1, the maximum DC-link voltage is 730 V. Contrarily, in region 2 the DC-link voltage will decrease to 600 V. The experimental results verify the theoretical analysis of Fig. 7.14. It can also be noted that the DC-link voltage control period is 8 ms.

After applying the coefficients autoregulation scheme in conventional method, the DC-link voltage can be controlled effectively when the d-axis current increases to the rated value as shown in Fig. 7.23. However, since the d-axis current is generated by PI regulator with the input of speed control error, the d-axis current adjustment is slow.

The experimental results of the proposed dual anti-overvoltage control scheme with controller coefficient autoregulation are shown in Fig. 7.24. $u_{dc\text{-}l1}$ is set at 640 V, and $u_{dc\text{-}l2}$ is set at 615 V. As can be seen in Fig. 7.24a, the DC-link voltage is controlled effectively during deceleration process, and the magnitude of the d-axis current can be controlled to the maximum value within 0.02 s benefited from the controller coefficient autoregulation. Consequently, the speed deceleration time can be reduced to 0.34 s from 45 to 5 Hz. Figure 7.24b is the zoomed view of the DC-link voltage and the dq-axis currents during active braking process. The maximum value of the voltage is 645 V, which is slightly higher than $u_{dc\text{-}l1}$ (640 V). Also, the DC-link voltage control period is reduced to 2.5 ms, which means the energy control error has been reduced effectively.

Figure 7.25 shows the experimental results when the rated load is added when

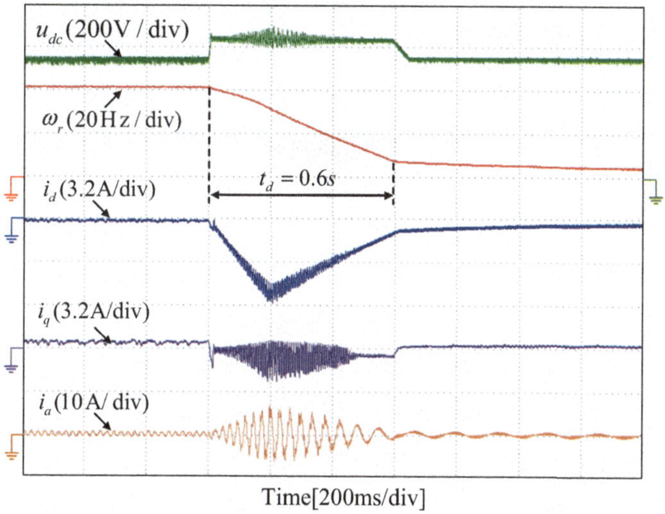

(a)

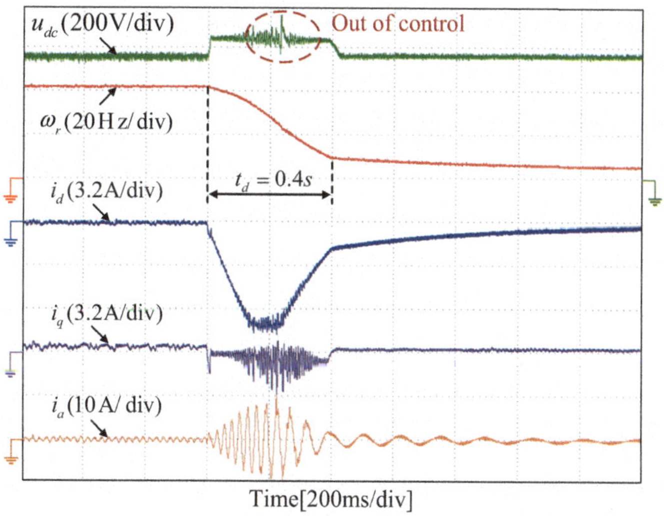

(b)

Fig. 7.22 Experimental results of speed decrease from 45 to 5 Hz using conventional anti-overvoltage scheme. **a** 80% d-axis current. **b** 100% d-axis current. **c** Influence of the q-axis current control delay

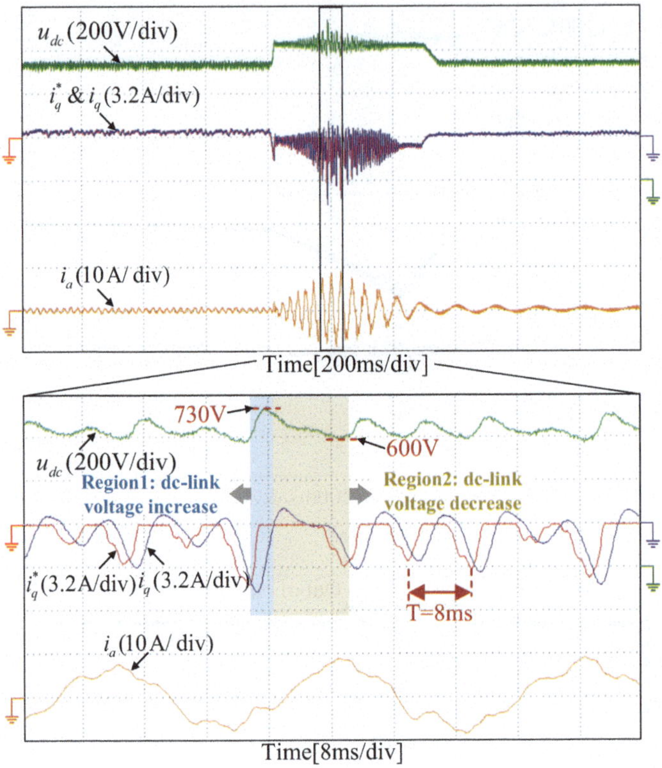

(c)

Fig. 7.22 (continued)

the speed is 45 Hz. The DC-link voltage increases slightly during speed deceleration process. The load torque consumes most of the regenerative energy, and the anti-overvoltage control methods have not activated. Figure 7.26 shows the experimental results when 50% rated load is added when motor speed is 45 Hz. Under this condition, the DC-link overvoltage phenomenon emerges. The DC-link voltage can be controlled effectively, and the speed deceleration time is 0.27 s using the conventional anti-overvoltage control scheme. When applying the proposed method, almost the same control performance is obtained, the speed deceleration time is 0.2 s. It should be noted that, the proposed anti-overvoltage control method works more effectively if the load torque is relatively large, since faster motor dynamic performance can be obtained.

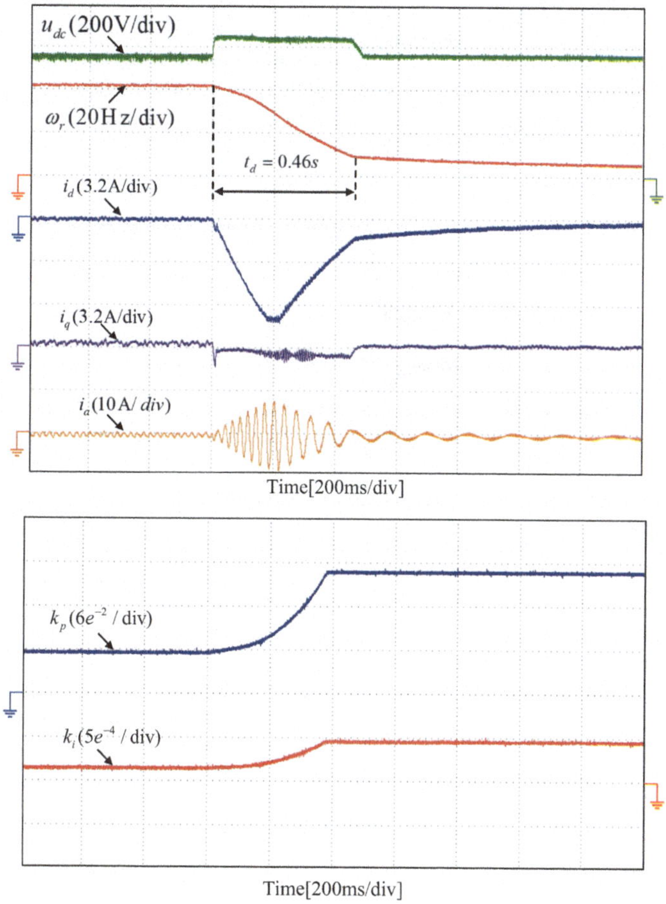

Fig. 7.23 Experimental results when motor speed decreases from 45 to 5 Hz. using conventional scheme with proposed autoregulated controller coefficients

The experimental results of the deceleration time statistics from different initial speeds to ultimate speeds are shown in Fig. 7.27. By using the proposed anti-overvoltage scheme, the deceleration time can be reduced effectively compared with the method in [4]. Specially, the deceleration time reduction is more obvious in lower speed region. Quantificationally, when the speed changes from 15 to 5 Hz, the deceleration time can be reduced to 33%. The deceleration time can be reduced to 57% as the speed changes from 45 to 35 Hz.

Figure 7.28 illustrates that the DC-link overvoltage can be prevented when the load is suddenly reduced by adopting the proposed control scheme. Figure 7.28a

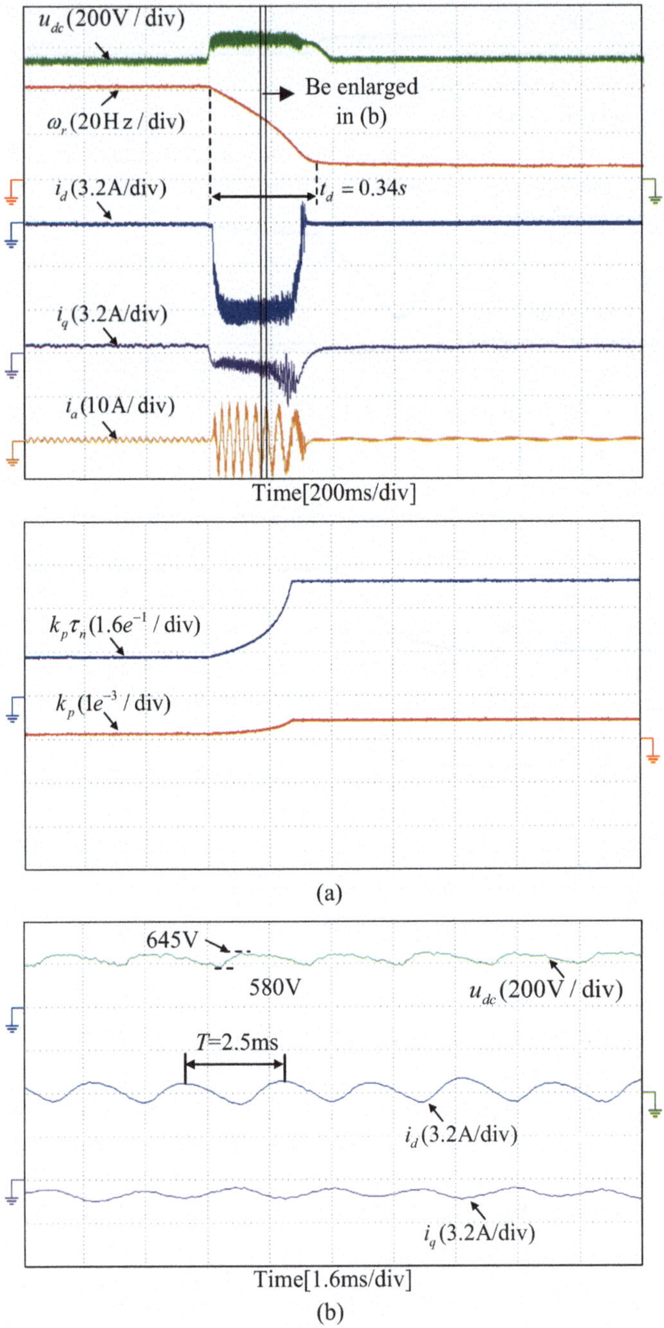

Fig. 7.24 Experimental results of speed decrease from 45 to 5 Hz using the proposed anti-overvoltage control scheme. **a** Whole braking process. **b** Zoomed view

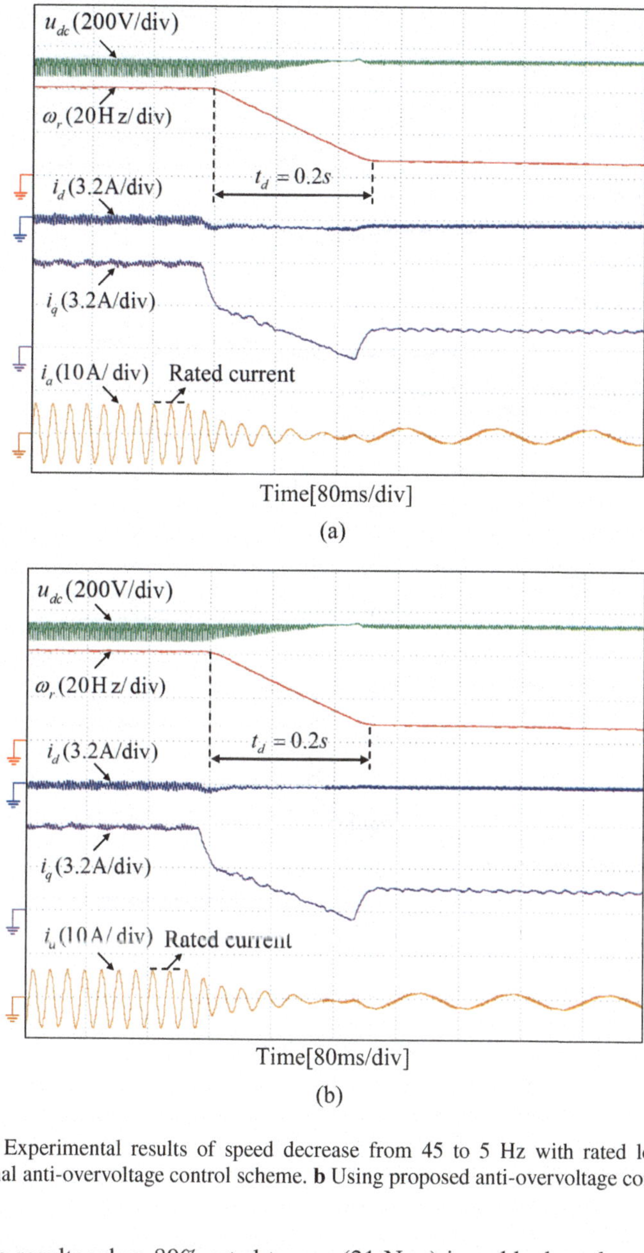

Fig. 7.25 Experimental results of speed decrease from 45 to 5 Hz with rated load. **a** Using conventional anti-overvoltage control scheme. **b** Using proposed anti-overvoltage control scheme

shows the results when 80% rated torque (21 N·m) is suddenly reduced to 0 N·m. During regenerative process, the DC-link voltage can be controlled effectively. The speed adjustment time is 0.35 s. Figure 7.28b shows that the DC-link voltage can still be controlled effectively when 100% rated torque (26 N·m) is suddenly reduced to 0 N·m.

Figure 7.29 shows the experimental results when different gaps between u_{dc_l1} and

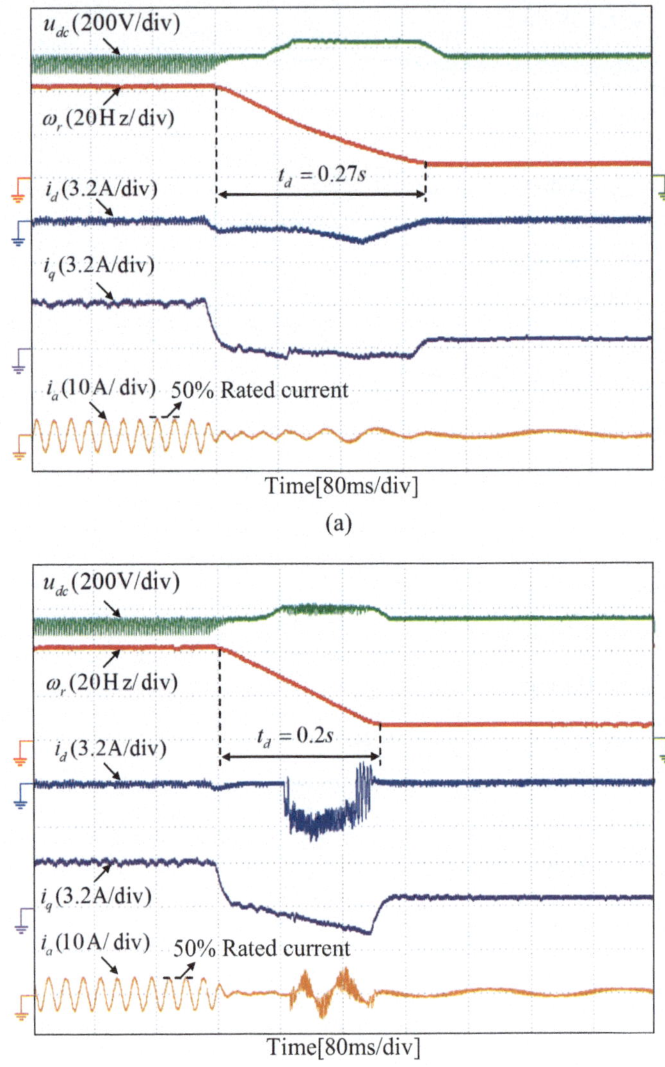

Fig. 7.26 Experimental results of speed decrease from 45 to 5 Hz with 50% load. **a** Using conventional anti-overvoltage control scheme. **b** Using proposed anti-overvoltage control scheme

u_{dc_l2} are set during speed deceleration process. From Fig. 7.29a, b, c, the voltage gap is increased. Along with the increase of voltage gap, the motor dynamic performance is improved. Specifically, when the speed decreases from 25 to 5 Hz, the speed adjustment times are 0.14 s and 0.11 s when voltage gaps are 10 V and 40 V, respectively. Although, the increase of voltage gap can improve the motor dynamic performance, the DC-link voltage control performance is degraded. As shown in

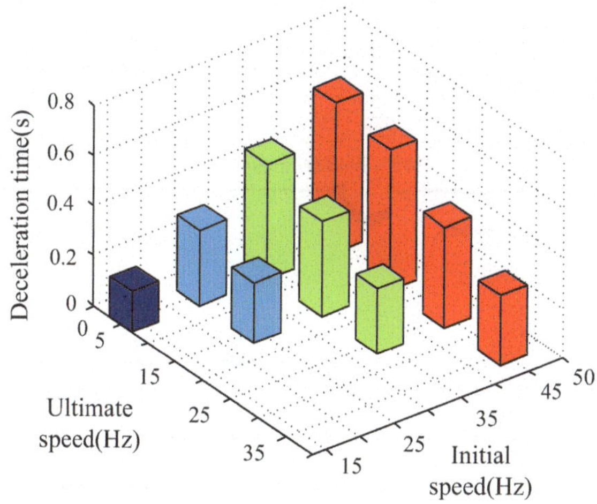

(a)

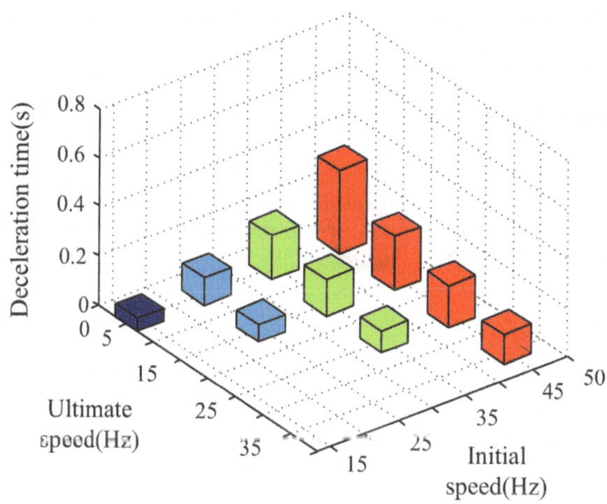

(b)

Fig. 7.27 Experimental statistical results of the deceleration time. **a** Conventional method. **b** Proposed method

Fig. 7.29c, the maximum DC-link voltage reaches 660 V. Figure 7.29d shows the experimental result when the motor speed decreases from 35 to 5 Hz using the voltage gap of 40 V, the DC-link voltage further increases to 690 V. Large voltage gap between u_{dc_l1} and u_{dc_l2} can improve the motor dynamic performance, but the DC-link voltage may be out of control. Small voltage gap may lead to slower motor dynamic performance. Thus, the settings of u_{dc_l1} and u_{dc_l2} should consider both motor dynamics and DC-link voltage control performance.

In order to verify the control performance under fluctuated AC input, the line-line voltage of the AC source produced by chroma 61512 is shown in Fig. 7.30. The

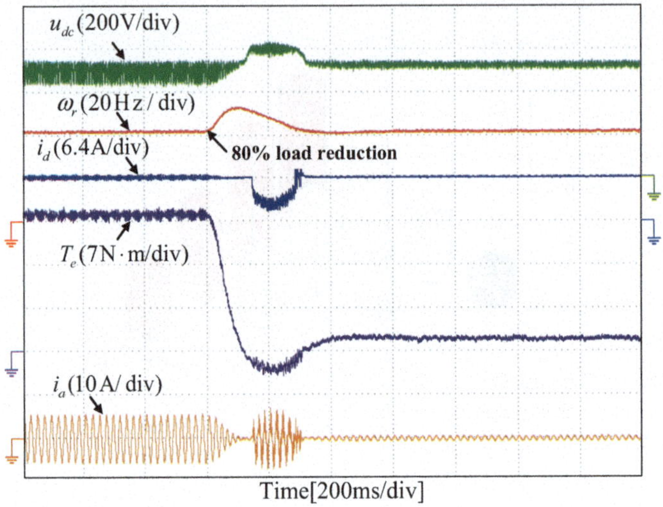

(a)

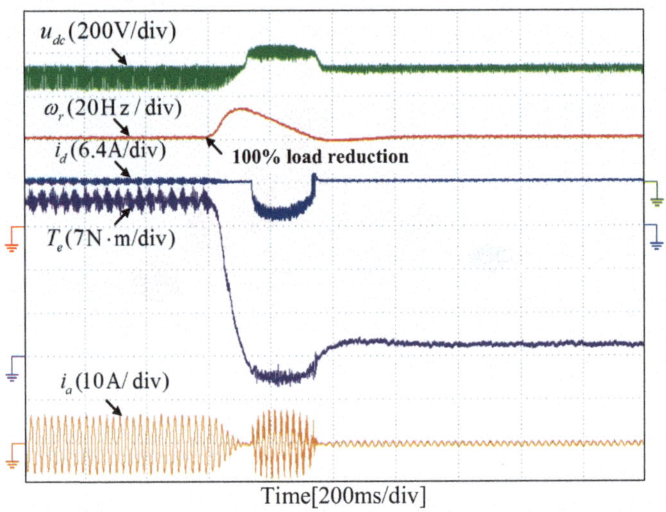

(b)

Fig. 7.28 Experimental results of load reduction at the speed of 45 Hz. **a** 80% load change. **b** 100% load change

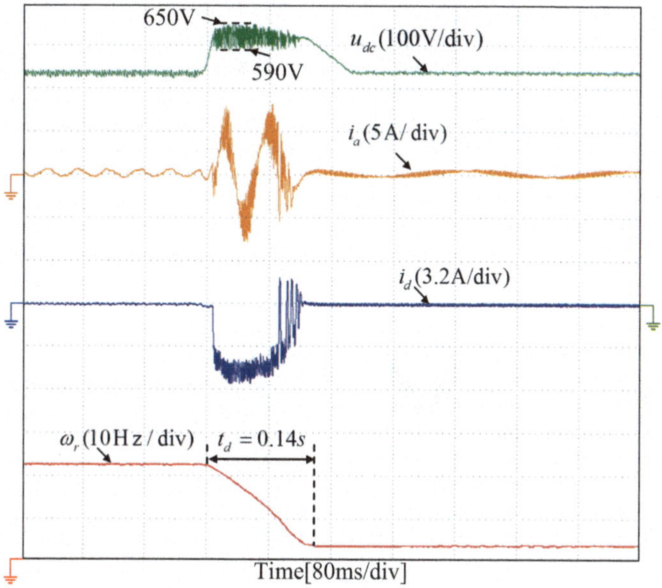

(a)

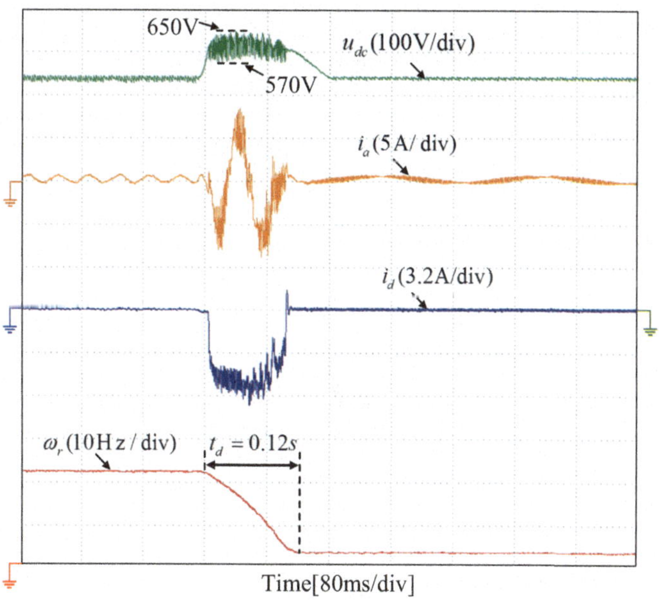

(b)

Fig. 7.29 Experimental results under different settings of u_{dc_l1} and u_{dc_l2}. **a** Speed decrease from 25 to 5 Hz when u_{dc_l1} = 640 V and u_{dc_l2} = 630 V. **b** Speed decrease from 25 to 5 Hz when u_{dc_l1} = 640 V and u_{dc_l2} = 615 V. **c** Speed decrease from 25 to 5 Hz when u_{dc_l1} = 640 V and u_{dc_l2} = 600 V. **d** Speed decrease from 35 to 5 Hz when u_{dc_l1} = 640 V and u_{dc_l2} = 600 V

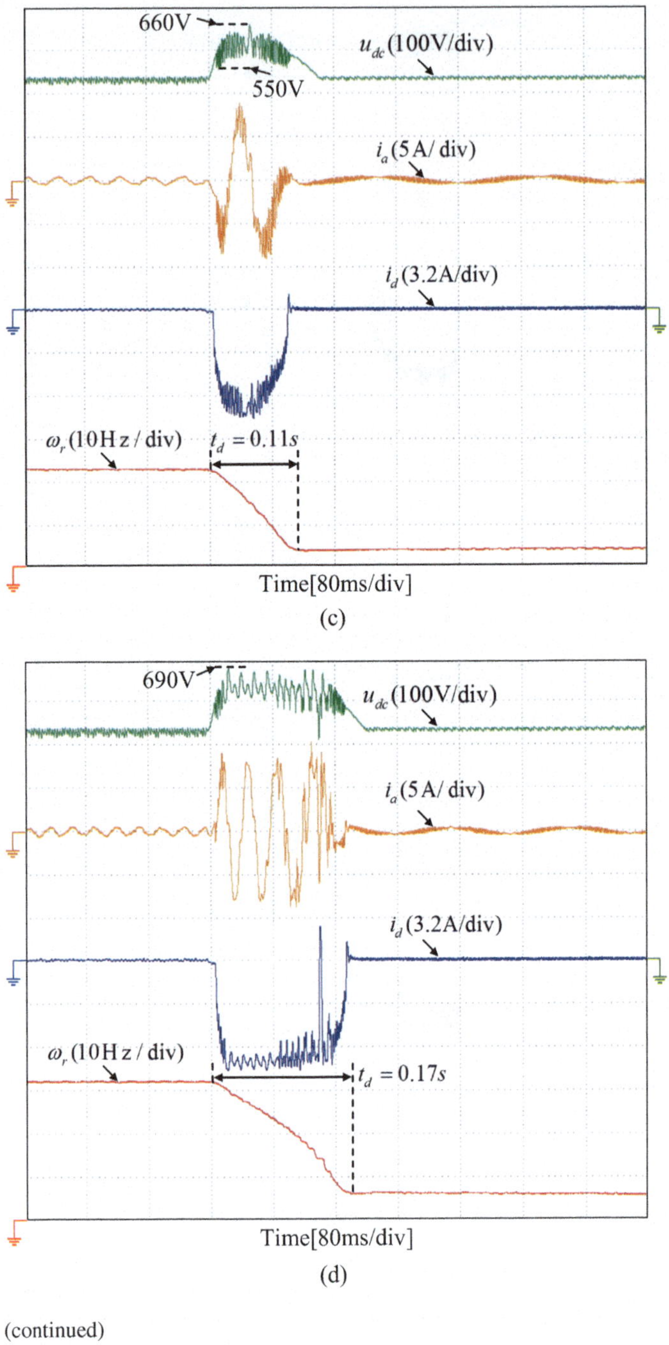

Fig. 7.29 (continued)

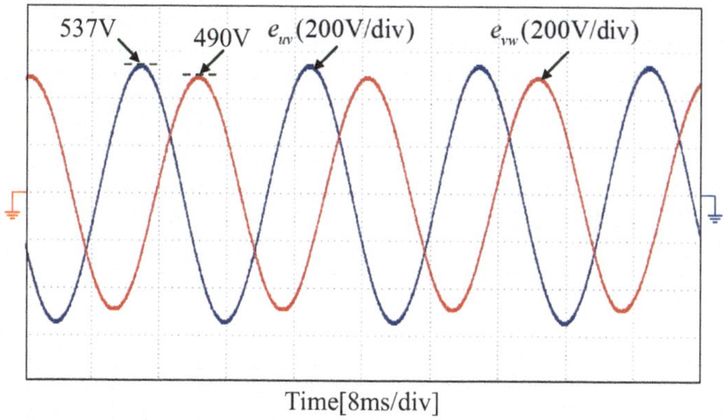

Fig. 7.30 Line-line voltage of AC source produced by chroma 61512

peak values of e_{uv} and e_{vw} are 537 V and 490 V, respectively. The frequency of the produced AC voltage is 50 Hz.

The anti-overvoltage control performance is tested under AC source fluctuation. Figure 7.31a shows the experimental results when speed deceleration under rated load. The DC-link voltage will increase slightly, and the speed deceleration time is 0.2 s. As can be seen, in region 1, the motor is operating in motoring mode with rated load. The peak value of the DC-link voltage is fluctuated which is affected by the AC input. In region 2, the motor is operating in regenerative braking mode, and the DC-link voltage is higher than the peak value of e_{uv} and e_{vw}. In region 3, the fluctuation of DC-link voltage is small since the motor power is largely reduced.

The speed deceleration without load is shown in Fig. 7.31b, the anti-overvoltage control performance is not affected by the AC source fluctuation. The DC link voltage can be controlled effectively, and the speed deceleration time is 0.33 s.

7.6 Summary

In this chapter, two anti-overvoltage control schemes are introduced for the electrolytic capacitorless motor drives during the motor transient process. Without additional hardware circuit, the DC-link voltage can be controlled effectively based on the motor loss concept. Experimental results show the effectiveness of the control methods.

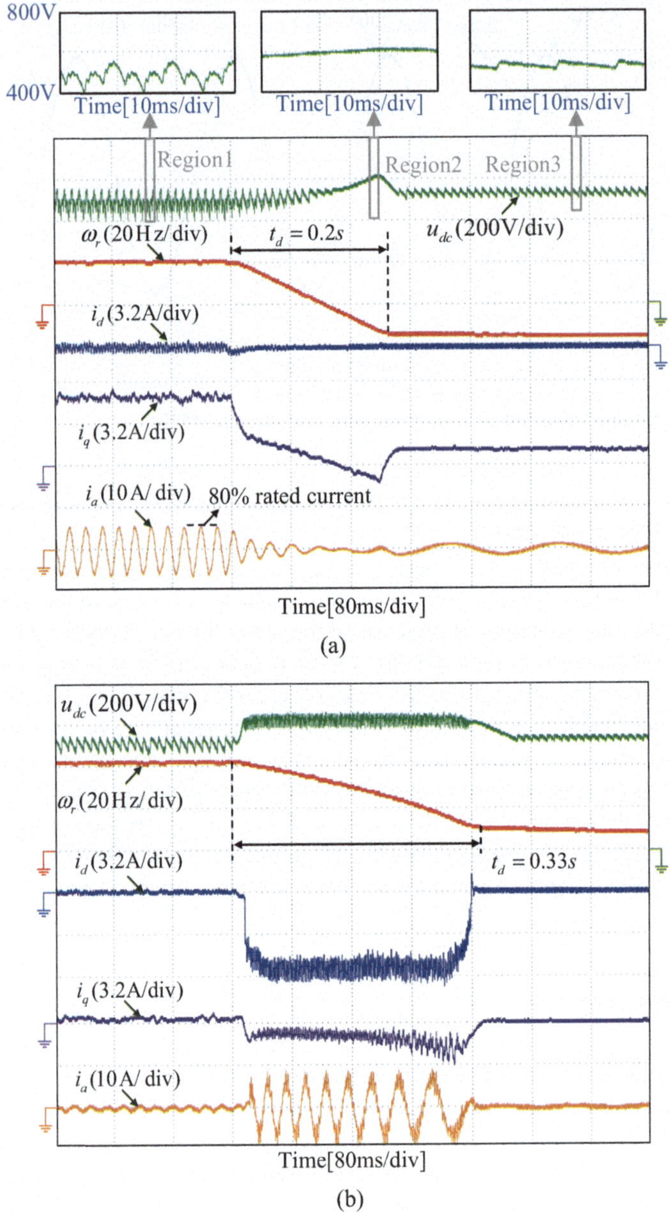

Fig. 7.31 Experimental results of speed decrease from 45 to 5 Hz under AC source fluctuation. **a** with rated load. **b** without load

References

1. R. Singh, S. Umashankar, D. Vijaykumar, D.P. Kothari, Dynamic braking of induction motor—analysis of conventional methods and an efficient multistage braking model, in *International Conference on Energy Efficient Technologies for Sustainability*, pp. 197–206 (2013)
2. Y. Wang, N. Niimura, R.D. Lorenz, Active braking schemes for low and high power induction machines using loss manipulation deadbeat-direct torque and flux control, in *European Conference on Power Electronics and Applications*, pp. 1–10 (2015)
3. M.M. Swamy, T. Kume, Y. Yukihira, F. Shuichi, M. Sawamura, A novel stopping method for induction motors operating from variable frequency drives. IEEE Trans. Power Electron. **19**(4), 1100–1107 (2004)
4. J. Jiang, J. Holtz, An efficient braking method for controlled AC drives with a diode rectifier front end. IEEE Trans. Ind. Appl. **37**(5), 1299–1307 (2001)
5. M. Rastogi, P.W. Hammond, Dual-frequency braking in AC drives. IEEE Trans. Power Electron. **17**(6), 1032–1040 (2002)
6. M. Hinkkanen, J. Luomi, Braking scheme for vector-controlled induction motor drives equipped with diode rectifier without braking resistor. IEEE Trans. Ind. Appl. **42**(5), 1257–1263 (2006)
7. K.P. Akinin, Research of braking mode of permanent magnet brushless motors without special brake devices, in *IEEE International Conference on Intelligent Energy and Power Systems*, pp. 257–260 (2014)
8. D. Ding, G. Zhang, G. Wang, D. Xu, Dual antiovervoltage control scheme for electrolytic capacitorless IPMSM drives with coefficient autoregulation. IEEE Trans. Power Electron, **35**(3), 2895–2907 (2020)
9. D. Ding, G. Wang, N. Zhao, G. Zhang, D. Xu, An antiovervoltage control scheme for electrolytic capacitorless IPMSM drives based on stator current vector orientation. IEEE Trans. Industr. Electron. **67**(5), 3517–3527 (2020)
10. S. Kim, Y. Yoon, S. Sul, K. Ide, Maximum Torque per Ampere (MTPA) control of an IPM machine based on signal injection considering inductance saturation. IEEE Trans. Power Electron. **28**(1), 488–497 (2013)
11. Y. Wang, T. Ito, R.D. Lorenz, Loss manipulation capabilities of deadbeat direct torque and flux control induction machine drives. IEEE Trans. Ind. Appl. **51**(6), 4554–4566 (2015)
12. M.J. Duran, I. Gonzalez-Prieto, F. Barrero, M. Mengoni, L. Zarri, E. Levi, A simple braking method for six-phase induction motor drives with diode front-end rectifier, in *Annual Conference of the IEEE Industrial Electronics Society*, pp. 1542–1547 (2015)
13. C. Kwan-Yuhl, Y. Soon-Bae, H. Chan-Hee, K. Jung-Chul, An efficient braking scheme for PM synchronous motor drives, in *International Conference on Electrical Machines and Systems*, pp. 80–83 (2003)
14. M.J. Nam, K.Y. Cho, H.W. Kim, B.M. Han, H.S. Chung, Braking algorithm considering voltage limit condition for surface mounted PM synchronous motor, in *International Conference on Electrical Machines and Systems*, pp. 1443–1447 (2015)
15. S. Yang, J. Chen, Controlled dynamic braking for switched reluctance motor drives with a rectifier front end. IEEE Trans. Industr. Electron. **60**(11), 4913–4919 (2013)
16. L. Niu, D. Xu, M. Yang, X. Gui, Z. Liu, On-line inertia identification algorithm for PI parameters optimization in speed loop. IEEE Trans. Power Electron. **30**(2), 849–859 (2015)
17. B. Wang, C. Luo, Y. Yu, G. Wang, D. Xu, Antidisturbance speed control for induction machine drives using high-order fast terminal sliding-mode load torque observer. IEEE Trans. Power Electron. **33**(9), 7927–7937 (2018)
18. G. Wang, Z. Li, G. Zhang, Y. Yu, D. Xu, Quadrature PLL-based high-order sliding-mode observer for IPMSM sensorless control with online MTPA control strategy. IEEE Trans. Energy Convers. **28**(1), 214–224 (2013)

Chapter 8
Optimized Overmodulation Strategy

8.1 Overmodulation Method of SVPWM

8.1.1 Conventional Overmodulation of SVPWM

Overmodulation strategies can be used to increase the DC-link voltage utilization rate, which enhances the output torque ability. In practice, the motor can operate beyond linear modulation region, even enter the six-step operation by implementing effective overmodulation strategies. Generally, the classical overmodulation strategies can be classified into two categories [1]. One is the single-mode strategy [2–10], and the other is the dual-mode strategy [11–15].

Three typical single-mode overmodulation strategies are summarized and their characteristics are discussed. In [2–5], the minimum phase angle error overmodulation strategy was investigated. This method only modifies the phase angle of the voltage vector, which is relatively simple and easy to implement. However, it can only reach the maximum modulation ratio of 0.952, which could not fully utilize the DC-link voltage. Another strategy named the minimum amplitude error overmodulation strategy was proposed in [2–4, 6]. Compared with the minimum phase angle error overmodulation strategy, this kind of method can increase the maximum modulation ratio to infinitely close to 1. However, this method is relatively complicated. In an effort to simplify the computation, a single-mode strategy adopted in the whole modulation region was proposed by Bolognani and other researchers [3, 8–10]. This method can make full use of the DC-link voltage and has advantage of simple computation processing.

The single-mode overmodulation causes larger voltage distortion although it can achieve high modulation ratio with less computation. The dual-mode overmodulation strategy was proposed by J. Holtz and other researchers [10–12]. This method divides the overmodulation region into two parts: overmodulation region I and II. In overmodulation region I, only the amplitude of the voltage vector is changed. While both the amplitude and the phase angle are changed in overmodulation region II. This method can synthesize the advantages of the above single-mode overmodulation

© The Author(s), under exclusive license to Springer Nature Singapore Pte Ltd. 2021

G. Wang et al., *Reduced DC-link Capacitance AC Motor Drives*,
https://doi.org/10.1007/978-981-15-8566-1_8

strategies, which ensures that the stator voltage vector is continuously transmitted from linear modulation region to the six-step operation. However, this method has heavy computing burden and occupies large memory resources. For solving this problem, a dual-mode overmodulation strategy was proposed in [13]. Compared with classical strategy, the method can simplify the computation processing. In [14], the linearity between the output voltage and the modulation ratio is guaranteed, which ensures the seamless transition from linear modulation region to the six-step operation.

8.1.2 Analysis of the Overmodulation in Reduced DC-Link Capacitance PMSM Drives

The above overmodulation strategies were implemented based on large-volume electrolytic capacitor drives, in which the DC-link voltage almost maintains a constant value. However, in electrolytic capacitorless motor drives, the DC-link voltage fluctuates seriously, which results in unintentional entrance to overmodulation region and affects the operation performance in overmodulation region. Consequently, the more effective overmodulation strategy is needed to improve the performance of reduced DC-link capacitance PMSM drives. The relation among the THD, the value of the fundamental voltage, the operation frequency and the initial phase of the voltage vector are analyzed in [15] when small DC-link capacitors are equipped. However, the overmodulation strategy proposed in this reference causes the phase-angle to jump frequently, which results in the distortion of the stator voltage and current.

In this section, the influence of the dual-mode overmodulation strategy for reduced DC-link capacitance PMSM drives is analyzed. According to the analysis, it can be concluded that the time for entering the six-step operation is difficult to control due to the comprehensive effect of the fluctuating DC-link voltage and the conventional dual-mode overmodulation strategy, which results in the increase of THD of the stator voltage and current. For solving this problem, a novel overmodulation strategy named optimized voltage boundary based overmodulation strategy is proposed. This strategy can improve the performance in overmodulation regions by switching between the actual DC-link voltage and the fixed DC-link voltage used for SVPWM. The fixed DC-link voltage is used for SVPWM in overmodulation regions, which can exactly control the time for entering the six-step operation. Besides, the selection of the optimal value of the fixed DC-link voltage is discussed.

8.2 Voltage Distortion Caused by Convensional Dual-Mode Overmodulation

The influence of DC-link voltage fluctuation on voltage boundary hexagon is shown as Fig. 8.1. As can be seen from Fig. 8.1, when the DC-link voltage decreases to the minimum value, it corresponds to the minimum voltage boundary hexagon. When the DC-link voltage increases to the maximum value, it corresponds to maximum voltage boundary hexagon. The size of the voltage boundary hexagon varies with the voltage fluctuation on the DC-link voltage. As can be seen from Fig. 8.1a, when the DC-link voltage increases, the voltage boundary hexagon will expand. As can be seen from Fig. 8.1b, the voltage boundary hexagon will shrink when the DC-link voltage reduces. The voltage boundary hexagon changes with the voltage fluctuation of the DC-link, which leads to extra harmonics when the motor operates in overmodulation region, affecting the operation performance of the motor drive.

When the motor operates in overmodulation region, the αβ-axes voltage in stationary frame will distort sharply. Figure 8.2 shows the voltage vector trajectory and waveforms when the motor drive operates in overmodulation region. It can be seen that two types of the waveform distortions occur while the conventional dual-mode overmodulation strategy is used for reduced DC-link capacitance PMSM drives. This Chapter classifies these distortions into two types: Type-I distortion and Type-II distortion.

Figure 8.2 shows the trajectory and αβ-axes voltages waveforms, where $\boldsymbol{u}_s^*$ is the reference stator voltage vector, $\boldsymbol{u}_s$ is the stator voltage vector, $\boldsymbol{u}_1$-$\boldsymbol{u}_6$ are the active vectors, u_α is the α-axis voltage, and u_β is the β-axis voltage. As can be seen from Fig. 8.2, ①②③④ is the trajectory of the Type-I distortion. When $\boldsymbol{u}_s$ moves from ③ to ④ in Fig. 8.2a, i.e. from D to E in Fig. 8.2b, the phase-angle of $\boldsymbol{u}_s$ jumps and reverses. This is since the voltage hexagon will expand and shrink repeatedly due to

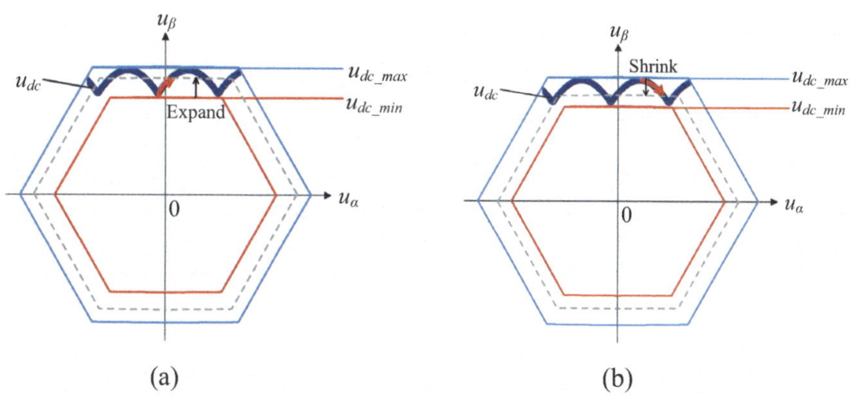

Fig. 8.1 Influence of DC-link voltage fluctuation on voltage boundary hexagon. **a** When DC-link voltage increases. **b** When DC-link voltage decreases

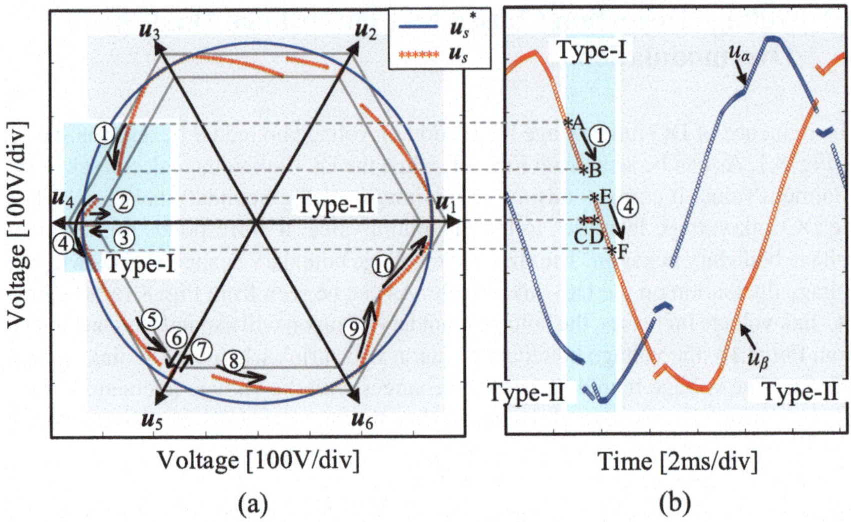

Fig. 8.2 Two types of stator voltage distortion. **a** trajectory. **b** αβ-axes voltage waveforms

the DC-link voltage fluctuation. The motor will exit from the six-step operation and return to the previous operating point owing to the expansion of the voltage hexagon. In addition, Type-II distortion exists and lasts for the whole overmodulation region. When $\boldsymbol{u}_s$ moves alone the edge of the hexagon or stays still in one of the hexagon vertexes, the value of the stator voltage vector $|\boldsymbol{u}_s|$ will change due to the expansion and contraction of voltage hexagon.

The expression of the DC-link voltage $u_{dc}(t)$can be expressed as

$$u_{dc}(t) = u_{dc\_max}\sin\left(\omega_g t + k\pi/3\right) \tag{8.1}$$

where $k = 1, 2, \ldots 6$ during one period.

The analysis of these two types of distortion is presented as follows:

(1) Type-I distortion

Here, the movement during sector III is considered as shown in Figs. 8.3 and 8.4. As can be seen from Fig. 8.3, when the conventional overmodulation strategy is used in general purpose PMSM drives using large-volume electrolytic capacitors, the DC-link voltage can be considered constant. Then the hexagon of the voltage vector is also constant. In this case, the phase-angle of the stator voltage vector $\boldsymbol{u}_s$ will not reverses.

As can be seen from Fig. 8.4, when the conventional dual-mode overmodulation strategy is used in reduced DC-link capacitance PMSM drives, the conventional dual-mode overmodulation determines whether the motor enters the six-step operation according to the comparison between $T_3/2 + T_4$ and T_s. Note that T_3 and T_4 are the action time of the two fundamental voltage vectors in the sector III, and T_s is

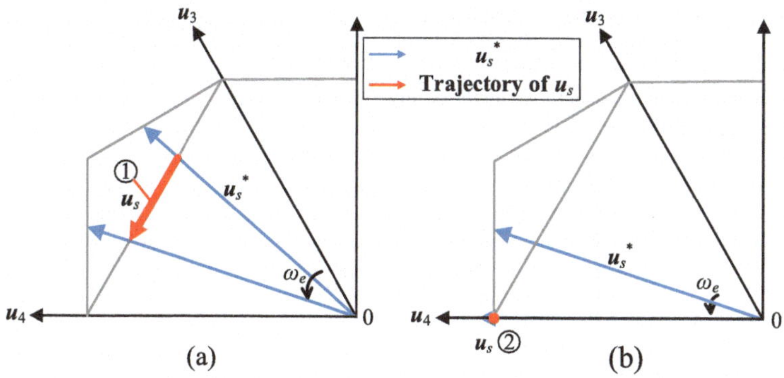

Fig. 8.3 Reference and stator voltage vector trajectories in general purpose PMSM drives. **a** Trajectory ①. **b** Trajectory ②

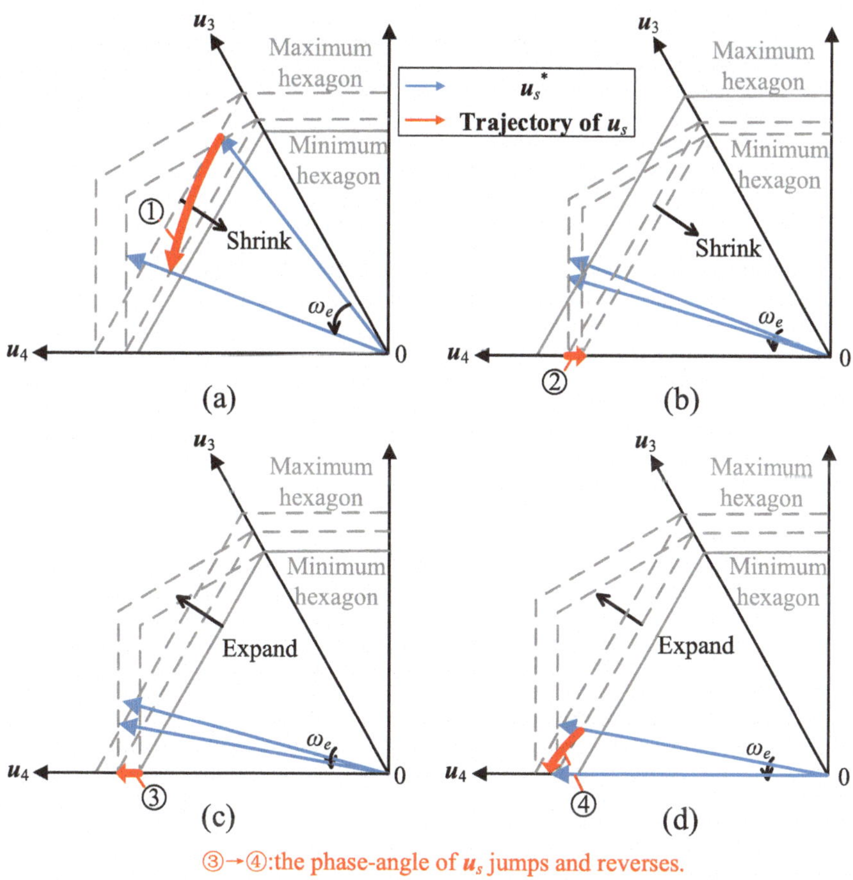

Fig. 8.4 Reference and stator voltage vector trajectories of Type-I distortion. **a** Trajectory ①. **b** Trajectory ②. **c** Trajectory ③. **d** Trajectory ④

the PWM carrier period. In reduced DC-link capacitance PMSM drives, the value of $T_3/2 + T_4$ is related to the DC-link voltage fluctuation.

The action time T_i and T_{i+1} of the fundamental voltage vectors in arbitrary sector can be expressed as

$$\begin{cases} T_i = \dfrac{\sqrt{3}T_s\left|\boldsymbol{u}_s^*\right|\sin\left(\pi/3 - \theta_{us}(t)\right)}{u_{dc}(t)} \\ T_{i+1} = \dfrac{\sqrt{3}T_s\left|\boldsymbol{u}_s^*\right|\sin(\theta_{us}(t))}{u_{dc}(t)} \end{cases} \tag{8.2}$$

where $\theta_{us}(t)$ is the angle between $\boldsymbol{u}_s^*$ and active vector $\boldsymbol{u}_i$, and the subscript "i" represents the sector number.

For sector III, substituting (8.1) into (8.2) can obtain the result as

$$T_3/2 + T_4 = \frac{\sqrt{3}T_s\left|\boldsymbol{u}_s^*\right|\left(\sin\left(\pi/3 - \theta_{us}(t)\right)/2 + \sin(\theta_{us}(t))\right)}{u_{dc}(t)}. \tag{8.3}$$

The numerator of (8.3) is defined as a function,

$$F(t) = \sqrt{3}T_s\left|\boldsymbol{u}_s^*\right|\left(\sin\left(\pi/3 - \theta_{us}(t)\right)/2 + \sin(\theta_{us}(t))\right). \tag{8.4}$$

The analysis of the Type-I distortion by combining the expression of $T_3/2 + T_4$ with the reference and the stator voltage vector trajectories shown in Fig. 8.4 is given as follows:

When $\boldsymbol{u}_s$ moves alone ① as shown in Fig. 8.4a, T_3 and T_4 satisfy $T_3 + T_4 > T_s$, and $\boldsymbol{u}_s$ moves alone the edge of the hexagon. T_3 and T_4 should be reduced to T_3' and T_4', respectively.

$$\begin{cases} T_3' = T_s \dfrac{T_3}{T_3 + T_4} \\ T_4' = T_s \dfrac{T_4}{T_3 + T_4} \end{cases}. \tag{8.5}$$

The trajectory ② is shown in Fig. 8.4b. At the beginning of the trajectory, $\boldsymbol{u}_s^*$ is on the boundary of the dotted line and $T_3/2 + T_4 = T_s$, which means the motor enters the six-step operation and $\boldsymbol{u}_s$ stays still in one of the hexagon vertexes. During the movement, the actual voltage hexagon draws back towards the minimum voltage hexagon, which means $du_{dc}/dt < 0$, and $dF/dt > 0$. Consequently, $T_3/2 + T_4$ keeps increasing and always satisfies $T_3/2 + T_4 > T_s$, i.e. the motor operates in a six-step operation state, which means $T_3 = 0$ and $T_4 = T_s$.

The trajectory ③ is shown in Fig. 8.4c. During the movement, the actual voltage hexagon expands towards the maximum voltage hexagon, which means that $du_{dc}/dt > dF/dt > 0$. Consequently, $T_3/2 + T_4$ will constantly decrease until it is equal to T_s.

The trajectory ④ is shown in Fig. 8.4d. During the movement, the actual voltage hexagon expands towards the maximum voltage hexagon, which means that $du_{dc}/dt > dF/dt > 0$. Consequently, $T_3/2 + T_4$ will constantly decrease until $T_3 + T_4 < T_s$, which means $\boldsymbol{u}_s$ exits from the six-step operation and returns to the previous operating point even to the linear modulation region, i.e. the phase-angle of $\boldsymbol{u}_s$ jumps and reverses.

(2) Type-II distortion

Since the Type-II distortion exists and lasts for the whole overmodulation region, the Type-II distortion can be analyzed in different overmodulation regions. When $\boldsymbol{u}_s$ moves alone ⑥⑦, the motor enters the six-step operation, which means $T_4 = 0$ and $T_5 = T_s$. The stator voltage vector $\boldsymbol{u}_s$ can be derived as

$$\boldsymbol{u}_s = \frac{T_4}{T_s}\boldsymbol{u}_4 + \frac{T_5}{T_s}\boldsymbol{u}_5 = \boldsymbol{u}_5. \tag{8.6}$$

The amplitude of the fundamental voltage vector fluctuates with the DC-link voltage fluctuation, which can be expressed as

$$\boldsymbol{u}_i = 2/3u_{dc}(t)e^{j\frac{\pi}{3}(i-1)}. \tag{8.7}$$

Substituting (8.7) into (8.6) when i is selected as 5, the following can be obtained

$$\boldsymbol{u}_s = 2/3u_{dc}(t)e^{j\frac{4\pi}{3}} \tag{8.8}$$

As can be seen in (8.8), the amplitude of the stator voltage vector $|\boldsymbol{u}_s|$ will fluctuate with the DC-link voltage fluctuation.

When $\boldsymbol{u}_s$ moves alone ⑨⑩, i.e. the edge of the hexagon, T_6 and T_1 should be reduced to T_6' and T_1', respectively.

$$\begin{cases} T_6^{'} = T_s\dfrac{T_6}{T_6 + T_1} \\ T_1^{'} = T_s\dfrac{T_1}{T_6 + T_1} \end{cases}. \tag{8.9}$$

The stator voltage vector can be derived as

$$\boldsymbol{u}_s = \frac{T_6^{'}}{T_s}\boldsymbol{u}_6 + \frac{T_1^{'}}{T_s}\boldsymbol{u}_1. \tag{8.10}$$

Substituting (8.2) and (8.7) into (8.10) when i are selected as 1 and 6 respectively, the result can be obtained as

$$\boldsymbol{u}_s = \frac{2}{3}\frac{\sin(\pi/3 - \theta_{us}(t))e^{j\frac{5\pi}{3}} + \sin(\theta_{us}(t))}{\sin(\pi/3 - \theta_{us}(t)) + \sin(\theta_{us}(t))}u_{dc}(t). \tag{8.11}$$

As can be seen in (8.11), $|\boldsymbol{u}_s|$ will fluctuate with the DC-link voltage fluctuation. According to the above analysis, it can be noticed that the Type-I distortion is expected to be eliminated, but the Type-II distortion is inherent due to the DC-link voltage fluctuation. In order to eliminate the Type-I distortion, the proposed overmodulation strategy named optimized voltage boundary based overmodulation strategy will be discussed in next section.

8.3 Transition Analysis of Uncontrollable Modulation Region

According to the above analysis of the Type-I distortion, $\boldsymbol{u}_s$ moves alone three different trajectories in an electrical period, which means that the transition of modulation region is chaos. The uncontrollable modulation region transition conditions will result in entering the six-step operation in advance, which is described as follows,

The ideal modulation processed from linear modulation region to the six-step operation is that $\boldsymbol{u}_s$ moves alone from a circular trajectory to the hexagon trajectory with the increase of $|\boldsymbol{u}_s^*|$, and the motor does not enter the six-step operation until a complete hexagon trajectory has been moved alone during an electrical period. However, using the conventional dual-mode overmodulation strategy in reduced DC-link capacitance PMSM drives will result in entering the six-step operation in advance, i.e. $\boldsymbol{u}_s$ can not move alone a complete hexagon during an electrical period.

Figure 8.5 shows the trajectory at different values of $|\boldsymbol{u}_s^*|$ [16]. As can be seen, when $|\boldsymbol{u}_s^*|$ is smaller than $u_{dc\_min}/\sqrt{3}$, Us only moves alone the circle. When $u_{dc\_min}/\sqrt{3} \leq |\boldsymbol{u}_s^*| < 2/3\ u_{dc\_min}$, $\boldsymbol{u}_s$ moves alone the circle or the edge of the hexagon depending on the location of $\boldsymbol{u}_s^*$ and u_{dc}. When $|\boldsymbol{u}_s^*| \geq 2/3\ u_{dc\_min}$, the motor will enter the six-step operation at some point. Consequently, whatever $|\boldsymbol{u}_s^*|$ is, $\boldsymbol{u}_s$ can not move alone a complete hexagon during an electrical period.

From the analysis of entering the six-step operation in advance, it can be noticed that the time for entering the six-step operation can not be controlled exactly. The voltage vector trajectory in multiple periods while $|\boldsymbol{u}_s^*|$ beyond the maximum voltage hexagon is shown in Fig. 8.6. As can be seen, even as $|\boldsymbol{u}_s^*|$ is constant, the time for entering the six-step operation is different due to the DC-link voltage fluctuation, which reflects in the distribution of $\boldsymbol{u}_s$. The inverted triangle reflects that the motor prefers to enter the six-step operation early at low DC-link voltage, and enter the six-step operation late at high DC-link voltage. Voltage boundary hexagons of different sizes will lead to different angles of entering the six-step operation, that is, different times of entering the six-step operation. The time of entering the six-step operation is entirely dependent on changes of $u_{dc}(t)$, which is uncontrollable.

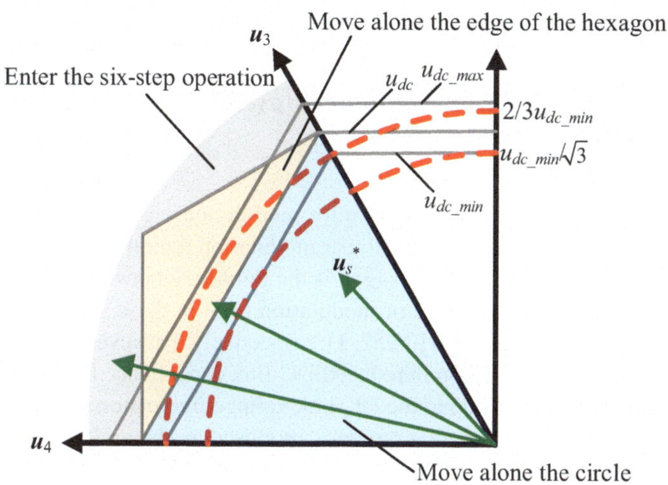

Fig. 8.5 Trajectory at different values of $|u_s^*|$

Fig. 8.6 Voltage vector trajectory in multiple periods

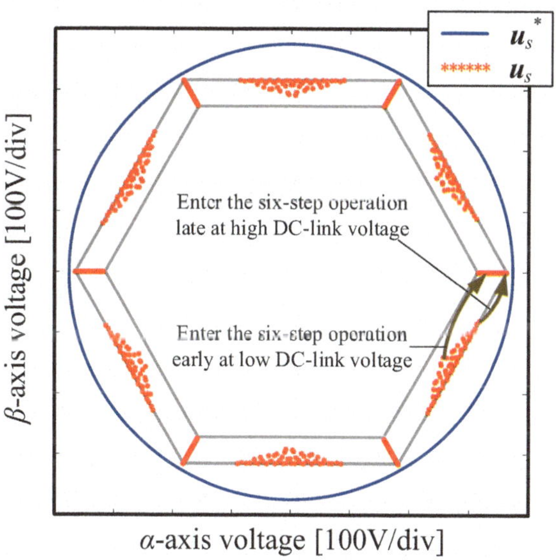

In conclusion, if the time for entering the six-step operation can be controlled exactly and the appropriate transition conditions can be designed, these problems can be solved. In order to control the time of entering the six-step operation, the influence of DC-link voltage fluctuation on T_i and T_{i+1} should be weakened so that the value of $T_i/2 + T_{i+1}(T_i + T_{i+1}/2)$ is controlled by u_s^*. For this purpose, optimized voltage boundary based overmodulation strategy is proposed.

8.4 Voltage Bundary Based Overmodulation Scheme

8.4.1 Optimized Voltage Boundary Based Overmodulation Strategy

According to the conventional definition for modulation region, the incircle of the hexagon is the boundary between the linear modulation region and the overmodulation region I. The excircle of the hexagon is the boundary between overmodulation region I and II. Besides, the range of modulation ratio in these modulation regions are [0, 0.907], [0.907, 0.952] and [0.952, 1], respectively. However, the conventional definition for modulation region in reduced DC-link capacitance PMSM drives will conflict and become chaos due to the DC-link voltage fluctuation.

A new definition of modulation region is proposed as shown in Fig. 8.7. As can be seen, the incircle of the actual voltage hexagon is the boundary between the linear modulation region and the overmodulation region I. While the excircle of the maximum voltage hexagon is the boundary between overmodulation region I and II.

For evaluating the modulation depth of the inverter in reduced DC-link capacitance PMSM drives, an alternative definition for modulation ratio is expressed as

$$M = \frac{|\boldsymbol{u}_s|\pi}{2RMS(u_{dc}(t))} \tag{8.12}$$

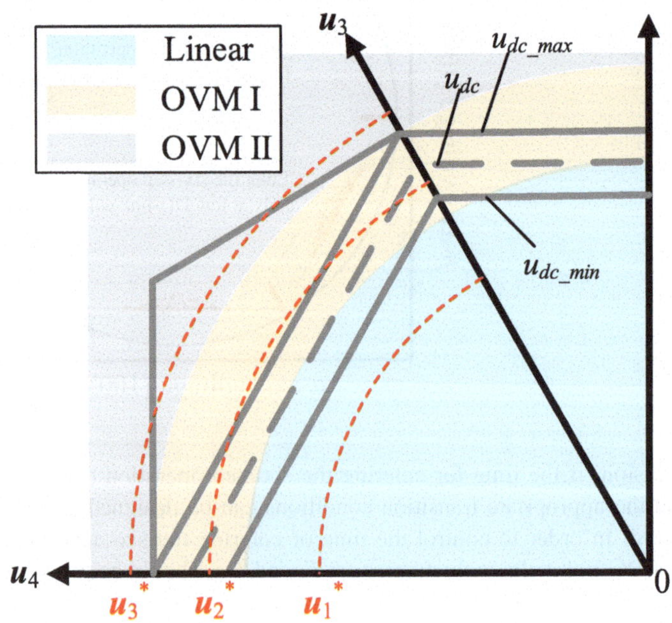

Fig. 8.7 Division of modulation region

where RMS ($u_{dc}(t)$) is the RMS value of the DC-link voltage.

Similarly, the distribution of the range of modulation ratio is the same as that in large-volume electrolytic capacitor drives, i.e. the range of modulation ratio in these modulation regions are also [0, 0.907], [0.907, 0.952] and [0.952, 1], respectively.

According to the above definition of modulation region, optimized voltage boundary based overmodulation strategy is introduced. The overmodulation strategy is implemented by switching between the actual DC-link voltage $u_{dc}(t)$ and the fixed DC-link voltage u_{fixed} used for SVPWM. The voltage switching time is based on the action time of the fundamental vectors when using u_{fixed} or $u_{dc}(t)$ for SVPWM, which can be expressed as

$$\begin{cases} T_{i\_fixed} = \dfrac{\sqrt{3}T_s\left|\boldsymbol{u}_s^*\right|\sin\left(\pi/3-\theta_{us}(t)\right)}{u_{fixed}} \\ T_{i+1\_fixed} = \dfrac{\sqrt{3}T_s\left|\boldsymbol{u}_s^*\right|\sin(\theta_{us}(t))}{u_{fixed}} \end{cases} \tag{8.13}$$

$$\begin{cases} T_{i\_actual} = \dfrac{\sqrt{3}T_s\left|\boldsymbol{u}_s^*\right|\sin\left(\pi/3-\theta_{us}(t)\right)}{u_{dc}(t)} \\ T_{i+1\_actual} = \dfrac{\sqrt{3}T_s\left|\boldsymbol{u}_s^*\right|\sin(\theta_{us}(t))}{u_{dc}(t)} \end{cases} \tag{8.14}$$

For the convenience of discussion, the following analysis is based on sector III, and the analysis of the other sectors can be obtained by similar methods.

(1) Linear modulation region ($0 \le M \le 0.907$)

As can be seen from Fig. 8.7, when reference stator voltage $\boldsymbol{u}_1^*$ moves in linear modulation region, $T_{3\_actual} + T_{4\_actual} \le T_s$. T_3 and T_4 can be expressed as

$$\begin{cases} T_3 = \dfrac{\sqrt{3}T_s\left|\boldsymbol{u}_s^*\right|\sin\left(\pi/3-\theta_{us}(t)\right)}{u_{dc}(t)} \\ T_4 = \dfrac{\sqrt{3}T_s\left|\boldsymbol{u}_s^*\right|\sin(\theta_{us}(t))}{u_{dc}(t)} \end{cases} \tag{8.15}$$

It is noticed that, $u_{dc}(t)$ is used in above formula, which ensures that $\boldsymbol{u}_s$ can completely follow the track of $\boldsymbol{u}_1^*$.

(2) Overmodulation region I ($0.907 < M \le 0.952$)

When $\boldsymbol{u}_2^*$ moves in OVM I, two modulation formulas is used depending on the location of $\boldsymbol{u}_2^*$.

When $\boldsymbol{u}_1^*$ moves in the actual voltage hexagon, $T_{3\_actual} + T_{4\_actual} \le T_s$. T_3 and T_4 can be expressed as (8.15).

When $\boldsymbol{u}_1^*$ moves outside the actual voltage hexagon, three inequalities satisfy simultaneously, i.e. $T_{3\_actual} + T_{4\_actual} > T_s$, $T_{3\_fixed}/2 + T_{4\_fixed} \le T_s$ and $T_{3\_fixed} +$

$T_{4\_fixed}/2 \le T_s$, and the minimum phase angle error overmodulation strategy is used to limit $\boldsymbol{u}_s$ to the actual voltage hexagon [2-5]. T_3 and T_4 can be expressed as

$$\begin{cases} T_3 = \dfrac{\sqrt{3}T_s|\boldsymbol{u}_s^*|\sin(\pi/3 - \theta_{us}(t))}{u_{fixed}} \\ T_4 = \dfrac{\sqrt{3}T_s|\boldsymbol{u}_s^*|\sin(\theta_{us}(t))}{u_{fixed}} \end{cases} \tag{8.16}$$

where T_3 and T_4 should be reduced to T_3' and T_4' as (8.5), respectively.

Actually, either $u_{dc}(t)$ or U_{fixed} is used in (8.16), $\boldsymbol{u}_s$ can be limited to the actual voltage hexagon due to the modified formula (8.5).

(3) Overmodulation region II ($0.952 < M \le 1$)

When $\boldsymbol{u}_3^*$ moves in OVM II, two modulation formulas are used depending on the location of $\boldsymbol{u}_3^*$.

When $\boldsymbol{u}_3^*$ moves in the triangle, $T_{3\_actual} + T_{4\_actual} > T_s$, $T_{3\_fixed}/2 + T_{4\_fixed} \le T_s$ and $T_{3\_fixed} + T_{4\_fixed}/2 \le T_s$, and the minimum phase angle error overmodulation strategy is used [2–5]. T_3' and T_4' can be expressed as (8.16) and (8.5).

When $\boldsymbol{u}_3^*$ moves outside the triangle, $T_{3\_fixed}/2 + T_{4\_fixed} > T_s$ or $T_{3\_fixed} + T_{4\_fixed}/2 > T_s$, and the motor enters the six-step operation, i.e. $\boldsymbol{u}_s$ is regulated to the nearest fundamental voltage vector. T_3 and T_4 can be expressed as

$$\begin{cases} T_3 = T_s \\ T_4 = 0 \end{cases} T_{3\_fixed} + T_{4\_fixed}/2 > T_s, \quad \begin{cases} T_3 = 0 \\ T_4 = T_s \end{cases} T_{3\_fixed}/2 + T_{4\_fixed} > T_s \tag{8.17}$$

From the discussion above, it can be concluded that T_{3_actual}, T_{4_actual}, T_{3_fixed} and T_{4_fixed} are used to determine which modulation formula should be used, which ensures that improved performance can be obtained in overmodulation region. The block diagram of the optimized voltage boundary based overmodulation strategy is shown in Fig. 8.8. The expressions corresponding to the modulation switching conditions in Fig. 8.8 are listed in Table 8.1.

When the optimized voltage boundary based overmodulation strategy is adopted, the selection of the optimal value of u_{fixed} should be well designed. Since different values of the u_{fixed} will result in different effects.

Figure 8.9 shows the comparison of M growth curve with three different values of u_{fixed}. As can be seen, when $u_{fixed1} < u_{dc\_max}$, M increases sharply at $2/3u_{fixed1}$, where the motor still operates in OVM I. That means the motor enters the six-step operation in advance. When $u_{fixed2} > u_{dc\_max}$, the gap between OVM I and OVM II occurs, and M is constant until $|\boldsymbol{u}_3^*| > 2/3u_{fixed2}$. In that case, the motor can not transit from OVM I to OVM II smoothly and swiftly. When $U_{fixed2} = u_{dc\_max}$, M increases smoothly to 0.952, and the gap between OVM I and OVM II will not occur. Consequently, u_{dc_max} is the optimal value of u_{fixed}.

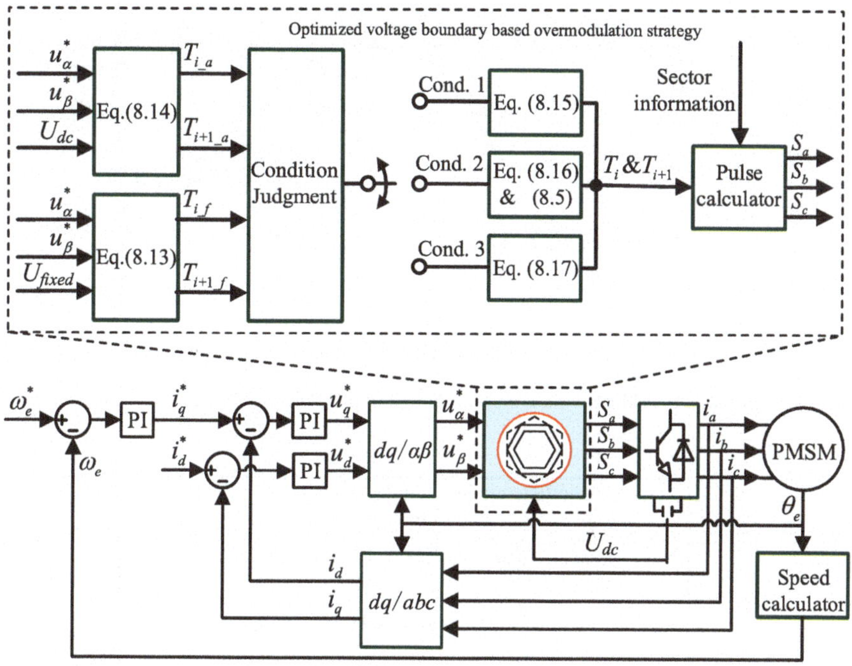

Fig. 8.8 Block diagram of optimized voltage boundary based overmodulation strategy

Table 8.1 SVPWM modulation strategy

Modulation switching condition	Expression
Cond. 1	$T_{i\_actual} + T_{i+1\_actual} \le T_s$
Cond. 2	$T_{i\_actual} + T_{i+1\_actual} > T_s$ & $T_{i\_fixed}/2 + T_{i+1\_fixed} < T_s$ & $T_{i\_fixed} + T_{i+1\_fixed}/2 \le T_s$
Cond. 3	$T_{i\_fixed}/2 + T_{i+1\_fixed} > T_s$ or $T_{i\_fixed} + T_{i+1\_fixed}/2 > T_s$

Figure 8.10 shows the curves of the drive efficiency and system efficiency using different overmodulation strategy. As can be seen, the drive efficiency and the system efficiency will increase with the motor power. The efficiency curves are not very different with different strategies. As a result, the drive efficiency is little influenced after using the proposed strategy.

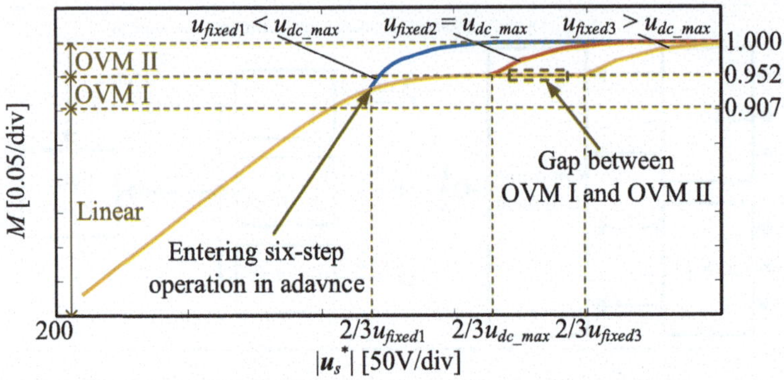

Fig. 8.9 Comparison of M growth curve with three different values of u_{fixed}

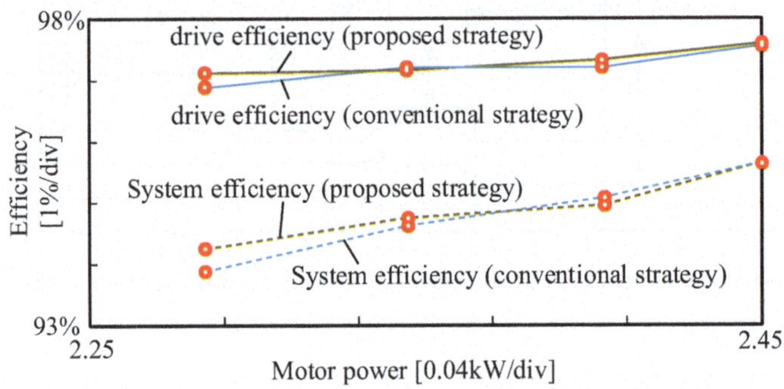

Fig. 8.10 Curves of the drive efficiency and system efficiency using different overmodulation strategies

8.4.2 Experimental Results of Optimized Voltage Boundary Based Overmodulation Strategy

The effectiveness of the voltage distortion analysis and the optimized voltage boundary based overmodulation strategy is verified experimentally. In the experiment, in order to satisfy the stability condition and minimize the switching ripple voltage, DC-link capacitance is set as 50 μF. A 2.2 kW PMSM drive with 50 μF film capacitors is used in the DC-link as shown in Fig. 6.13. When the motor operates in OVM II, the Type-I distortion occurs without using the optimized voltage boundary based overmodulation strategy is shown in Fig. 8.11a. As can be seen, the angle of the stator voltage vector θ_u falls back in the dotted line, which means that the phase-angle of $\boldsymbol{u}_s$ jumps and reverses. In this case, a spike of the α-axis voltage u_α appears, which results in the increase of THD and the distortion of the a-phase current i_a. By comparison with the results, the optimized voltage boundary based overmodulation strategy can eliminate this Type-I distortion as shown in Fig. 8.11b.

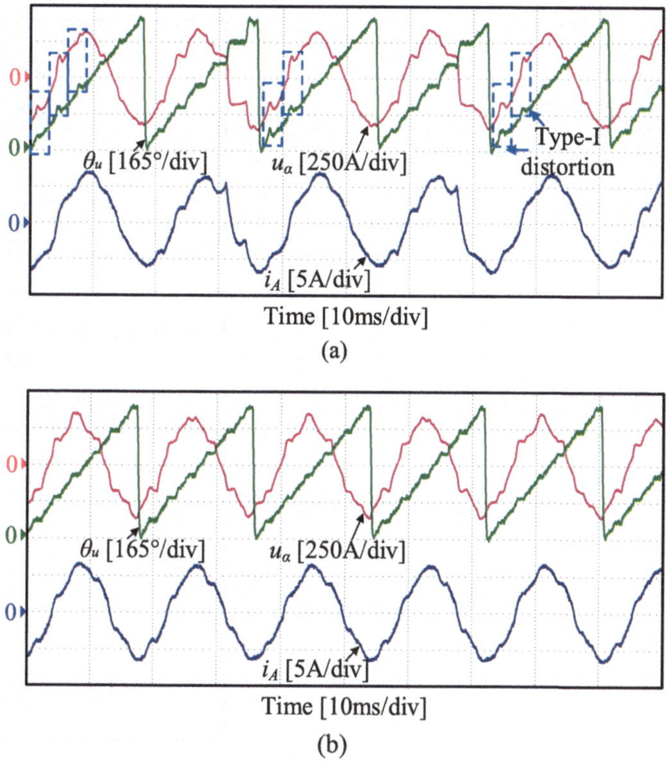

Fig. 8.11 Comparison of the Type-I distortion occurrence when motor operates in OVM II. **a** Without optimized voltage boundary based overmodulation strategy. **b** With optimized voltage boundary based overmodulation strategy

It is noticed that θ_u always increases in an electrical period by using the optimized voltage boundary based overmodulation strategy. Same conclusions can be obtained as shown in Fig. 8.12. As can be seen from Fig. 8.12a, the phase-angle of $\boldsymbol{u}_s$ jumps and reverses near the hexagon vertex in an electrical period. As a contrast, using the proposed overmodulation strategy can eliminate Type-I distortion as shown in Fig. 8.12b.

The FFT analyses of the i_a in the low frequency range are shown in Fig. 8.13. The result without using the optimized voltage boundary based overmodulation strategy is shown in Fig. 8.13a, i_a contains lots of harmonics including various low frequency harmonics. Besides, the THD of i_a reaches up to 19.3%. When the optimized voltage boundary based overmodulation strategy is used, the harmonics of i_a decreases is shown in Fig. 8.13b. The THD of i_a is decreased to 14.4%.

The FFT analysis of the phase current i_a around switching frequency range is shown in Fig. 8.14. The result without using the optimized voltage boundary based overmodulation strategy is shown in Fig. 8.14a, i_a contains some high frequency noises around switching frequency range (mainly at 6 kHz ± 0.055 kHz*2, 4,

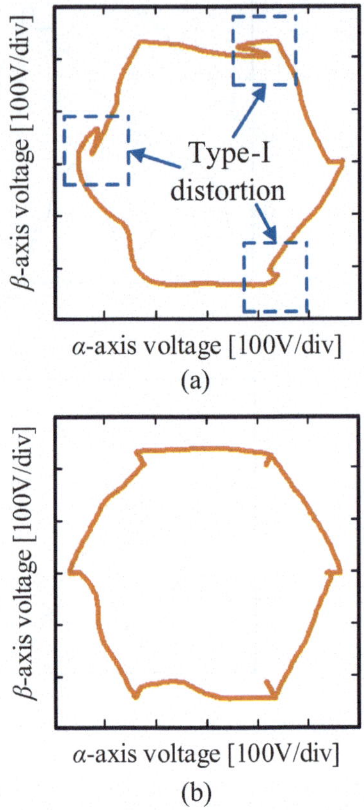

Fig. 8.12 Comparison of the Lissajous figures in an electrical period. **a** Without optimized voltage boundary based overmodulation strategy. **b** With optimized voltage boundary based overmodulation strategy

8, 10, where 0.055 kHz is the motor operating frequency). When the optimized voltage boundary based overmodulation strategy is used as shown in Fig. 8.14b, the switching frequency coupled with the motor operating frequency will increase slightly. However, other various frequency harmonics decrease, which means the harmonics around the switching frequency range are concentrated on the switching frequency coupled with the motor operating frequency (6 kHz $\pm$ 0.055 kHz*2, 4, 8, 10).

The FFT analysis of grid current is shown in Fig. 8.15. It can be seen that, after adopting the optimized voltage boundary based overmodulation strategy, the amplitude of 350 Hz harmonic increases slightly. The amplitude of 350 Hz harmonic increases from 1.16A to 1.49A. The amplitude of 250 Hz harmonic decreases slightly. The amplitude of 250 Hz harmonic decreases from 1.83A to 1.62A, and the amplitude of other frequency harmonics does not change much. In addition, after adopting optimized voltage boundary based overmodulation strategy, THD of the grid current decreases from 57.4% to 56.6%.

As can be seen from Fig. 8.16, in order to represent the modulation region that the motor operates at any time, a variable named modulation region flag FMR is

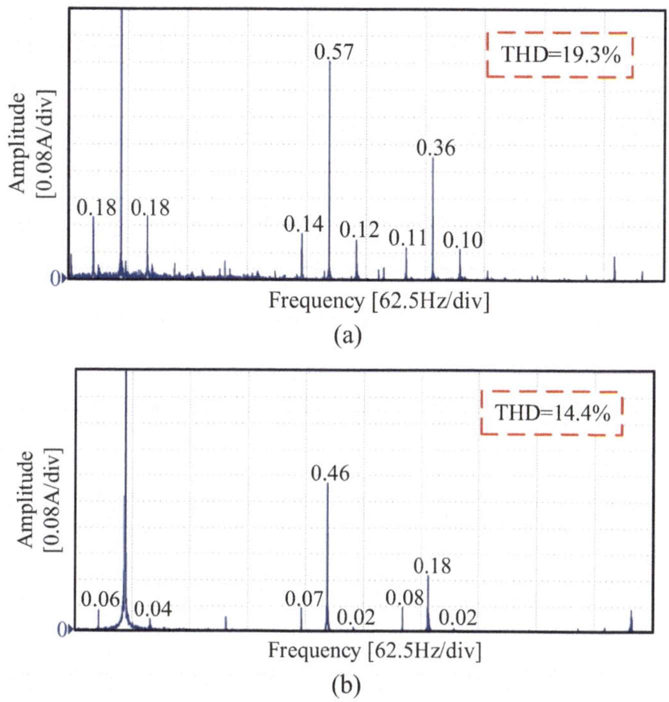

Fig. 8.13 Comparison of the FFT analyses of i_a in the low frequency range. **a** Without optimized voltage boundary based overmodulation strategy. **b** With optimized voltage boundary based overmodulation strategy

defined. This variable has three values of 0, 1 and 2, which represent $\boldsymbol{u}_s$ moves alone a circular trajectory, hexagon trajectory and enters the six-step operation, respectively. The transition of modulation region is chaos without using the optimized voltage boundary based overmodulation strategy as shown in Fig. 8.16a. The value of FMR changes among three different values during an electrical period, which means the motor enters the six-step operation in advance. The waveform distortion of u_α is different in various electrical periods. Besides, before the motor enters the six-step operation, a complete hexagon trajectory does not appear, which means the transition of modulation region is unideal. While the optimized voltage boundary based overmodulation strategy is used, the transition of modulation region is progressive as shown in Fig. 8.16b. The value of FMR increases gradually with the deepening of modulation depth, and a complete hexagon trajectory is moved alone before the motor enters the six-step operation. That means the motor can operate from linear modulation region to OVM II smoothly, and avoid entering the six-step operation in advance.

In an effort to evaluate the time for entering the six-step operation, an angle named α_h is defined as shown in Fig. 8.17. When the motor operates in OVM II, α_h is different or even decrease to zero without using the optimized voltage boundary

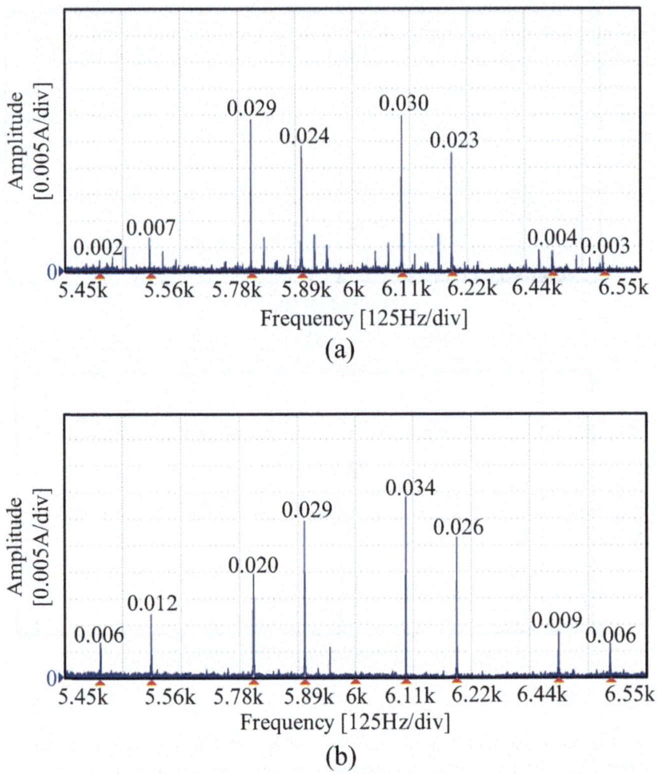

Fig. 8.14 Comparison of the FFT analysis of i_a around switching frequency range. **a** Without using the optimized voltage boundary based overmodulation strategy. **b** Using the optimized voltage boundary based overmodulation strategy

based overmodulation strategy as shown in Fig. 8.17a. While the optimized voltage boundary based overmodulation strategy is used, α_h is approximately equal during an electrical period as shown in Fig. 8.17b.

Same conclusions can be obtained as shown in Fig. 8.18. As can be seen from Fig. 8.18a, the trajectory of $\boldsymbol{u}_s$ represents as an inverted triangle, which means α_h is different or even decrease to 0 without using the optimized voltage boundary based overmodulation strategy. The trapezoid means α_h is approximately equal during an electrical period when the proposed overmodulation strategy is used as shown in Fig. 8.18b. The approximately identical α_h can reduce the degree of distortion of the u_α and decrease THD.

The comparison of experiments results is shown in Fig. 8.19. As can be seen, before the motor operates into overmodulation region, the waveforms of the torque T_e, the speed ω_r and the q-axis current i_q are the same respectively. When the motor operates in overmodulation region, the ripple of T_e, ω_r and i_q can be decreased from

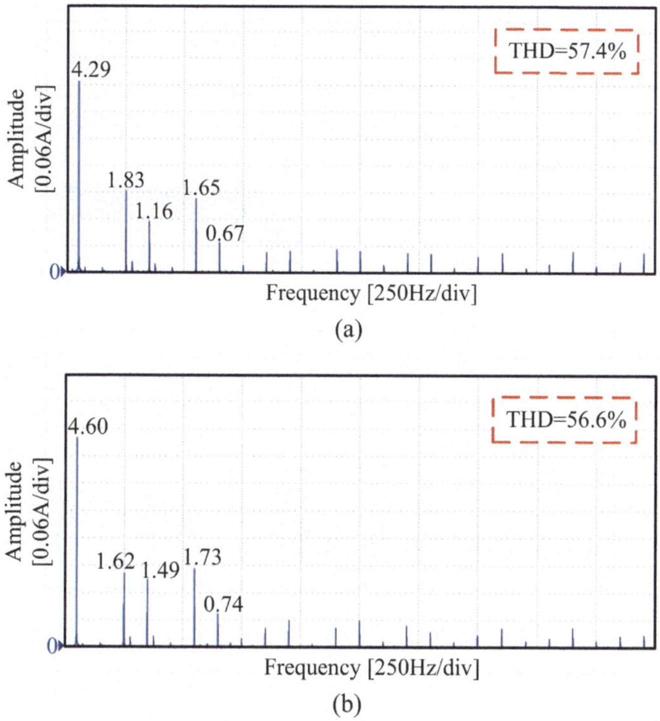

Fig. 8.15 Comparison of the FFT analysis of grid current. **a** Without using optimized voltage boundary based overmodulation strategy. **b** Using optimized voltage boundary based overmodulation strategy

26.31% to 14.04%, 2.00% to 1.14%, 24.50% to 12.68%, respectively by using the optimized voltage boundary based overmodulation strategy.

8.5 Summary

This Chapter focuses on the overmodulation strategy in three-phase diode rectifier PMSM drives equipped with slim film capacitors. The characteristics of the conventional dual-mode overmodulation strategy are analyzed, which includes the Type-I distortion, entering the six-step operation in advance and the difference of the time for entering the six-step operation. The optimized voltage boundary based overmodulation strategy has been proposed and the selection of optimal the value of the constant DC-link voltage for SVPWM has been discussed. The experimental results verify that the distortion of voltage and current can be improved when the optimized voltage boundary based overmodulation strategy is used.

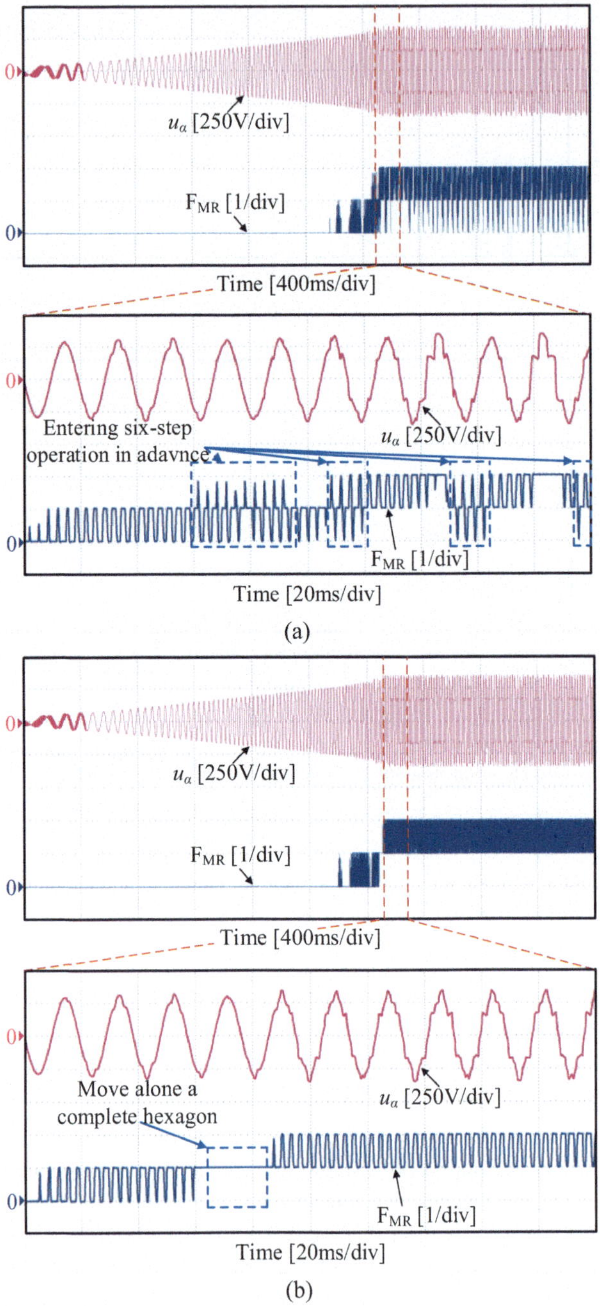

Fig. 8.16 Comparison of the transition of modulation region when motor operates from linear modulation region to OVM II. **a** Without optimized voltage boundary based overmodulation strategy. **b** With optimized voltage boundary based overmodulation strategy

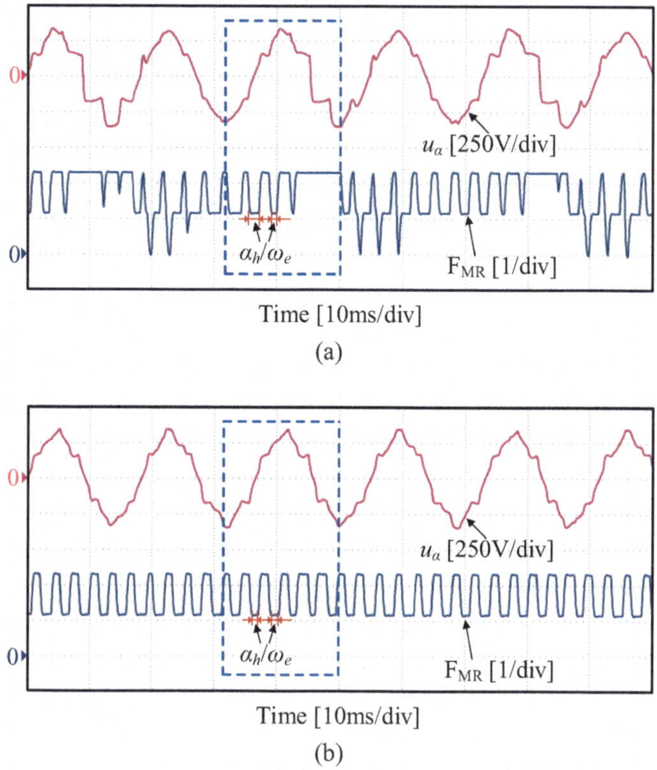

Fig. 8.17 Comparison of the angle α_h when the motor operates when motor operates in OVM II. **a** Without optimized voltage boundary based overmodulation strategy. **b** With optimized voltage boundary based overmodulation strategy

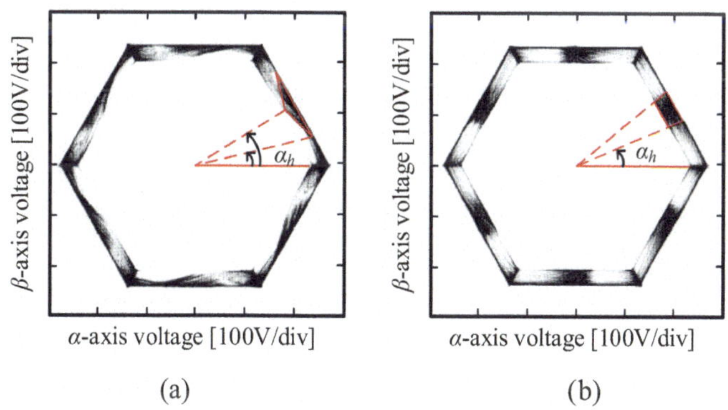

Fig. 8.18 Comparison of the Lissajous figures. **a** Without optimized voltage boundary based overmodulation strategy. **b** With optimized voltage boundary based overmodulation strategy

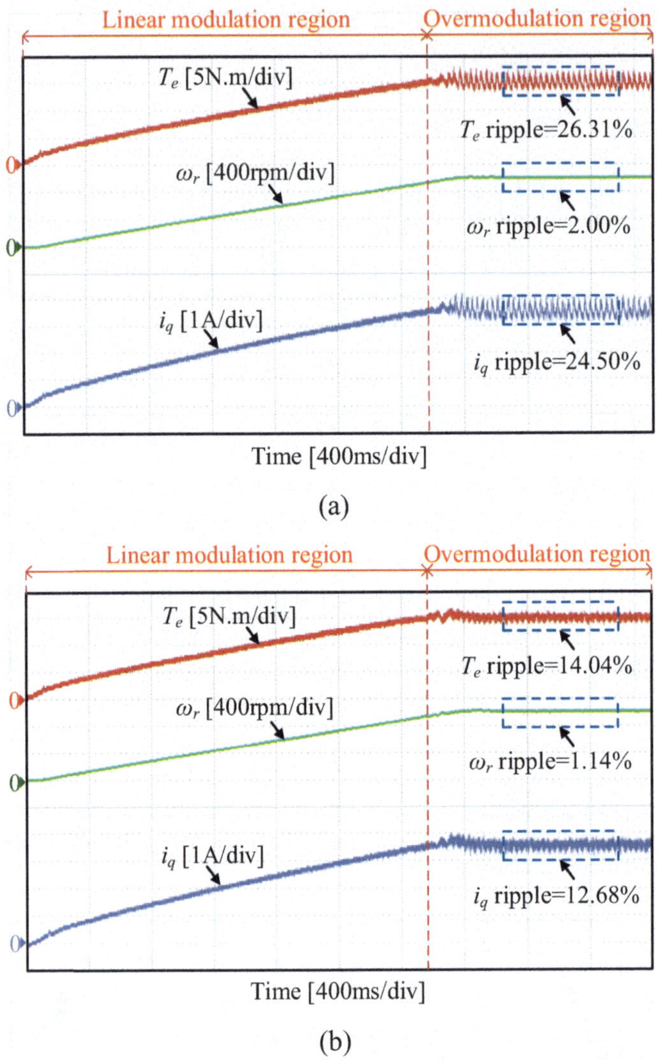

Fig. 8.19 Comparison of the experiments results when motor operates from linear modulation region to OVM II. **a** Without optimized voltage boundary based overmodulation strategy. **b** With optimized voltage boundary based overmodulation strategy

References

1. X. Guo, M. He, Y. Yang, Over modulation strategy of power converters: a review. IEEE Access **6**, 69528–69544 (2018)
2. F. Briz, A. Diez, M.W. Degner, R. D. Lorenz, Current and flux regulation in field-weakening operation [of induction motors]. IEEE Trans. Ind. Appl **37**(1), 42–50 (2001)

3. Z. Xu, D. Liu, X. Zhao, J. Ren, Over-modulation control strategy of SVPWM review, in *Chinese Control and Decision Conference* (2016), pp. 3192–3196
4. Y. Li, P. Dai, Y. Yu, X. Cao, Z. Jiang, Overview of SVPWM over modulation strategy of three-level inverter. Electric Drive **40**(7), 8–12 (2010)
5. T.G. Habetler, F. Profumo, M. Pastorelli, L.M. Tolbert, Direct torque control of induction machines using space vector modulation. IEEE Trans. Ind. Appl. **28**(5), 1045–1053 (1992)
6. D.R. Seidl, D.A. Kaiser, R.D. Lorenz, One-step optimal space vector PWM current regulation using a neural network, in *IEEE Industry Applications Society Annual Meeting* (1994), pp. 867–874
7. S. Bolognani, M. Zigliotto, Novel digital continuous control of SVM inverters in the overmodulation range. IEEE Trans. Ind. Appl. **33**(2), 525–530 (1997)
8. L. Zhang, J. Liu, X. Wen, G. Chen, Y. Hua, A novel algorithm of svpwm inverter in the overmodulation region based on fundamental voltage amplitude linear output control. Proc. Chin. Soc. Electr. Eng. **25**(19), 12–18 (2005)
9. Y. Kwon, S. Kim, S. Sul, Six-step operation of PMSM with instantaneous current control. IEEE Trans. Ind. Appl. **50**(4), 2614–2625 (2014)
10. G. Narayanan, V.T. Ranganathan, Extension of operation of space vector PWM strategies with low switching frequencies using different overmodulation algorithms. IEEE Trans. Power Electron. **17**(5), 788–798 (2002)
11. J. Holtz, W. Lotzkat, A.M. Khambadkone, On continuous control of PWM inverters in the overmodulation range including the six-step mode. IEEE Trans. Power Electron. **8**(4), 546–553 (1993)
12. D.C. Lee, G.M. Lee, A novel overmodulation technique for space-vector PWM inverters. IEEE Trans. Power Electron. **13**(6), 1144–1151 (1998)
13. X. Tang, X. Yang, S. Zhao, Flux analysis of one novel SVPWM overmodulation algorithm and its application in PMSM drive, in *International Conference on Electrical Machines and Systems* (2013), pp. 1166–1168
14. H. Lee, S. Hong, J. Choi, K. Nam, J. Kim, Sector-Based Analytic Overmodulation Method. IEEE Trans. Ind. Electron. **66**(10), 7624–7632 (2019)
15. H. Saren, O. Pyrhonen, K. Rauma, O. Laakkonen, Overmodulation in voltage source inverter with small DC-link capacitor, in *IEEE International Symposium on Industrial Electronics* (2005), pp. 892–898
16. G. Wang, H. Hu, D. Ding, N. Zhao, Y. Zou, D. Xu, Overmodulation strategy for electrolytic capacitorless PMSM drives: Voltage distortion analysis and boundary optimization. IEEE Trans. Power Electron. **35**(9), 9574–9585 (2020)

The manufacturer's authorised representative in the EU is Springer Nature Customer Service Centre GmbH, Europaplatz 3, 69115 Heidelberg, Germany. If you have any concerns regarding our products, please contact ProductSafety@springernature.com

Printed and bound by CPI Group (UK) Ltd, Croydon, CR0 4YY
24/07/2026
02175943-0002